Science and the Evolution of Consciousness

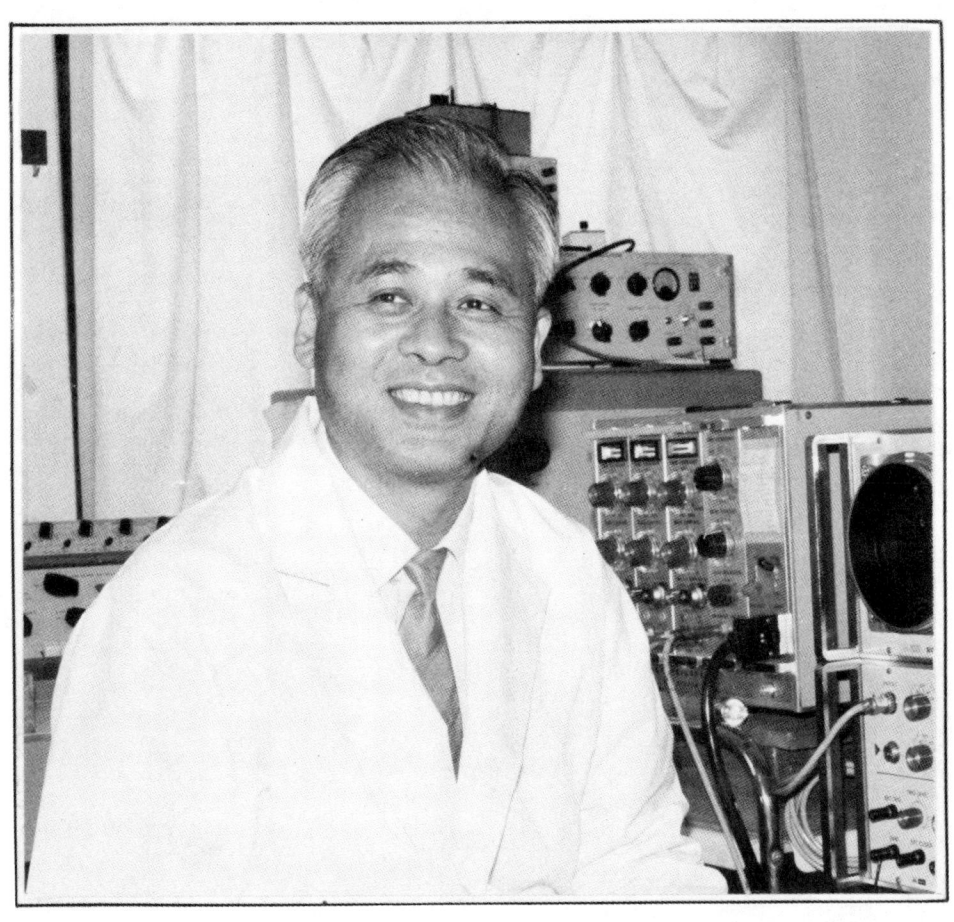

Dr. Hiroshi Motoyama

Science and the Evolution of Consciousness
Chakras, Ki, and Psi

Dr. Hiroshi Motoyama

with

Rande Brown

Autumn Press

Published by Autumn Press, Inc.
with editorial offices at
25 Dwight Street
Brookline, Massachusetts 02146
Distributed in the United States
by Random House, Inc. and in
Canada by Random House of Canada, Ltd.
Copyright © 1978 by Hiroshi Motoyama
All rights reserved.
Library of Congress Catalog Card Number: 78-51388
ISBN: 0-394-73634-6
Typeset at dnh typesetting, inc., Cambridge, Massachusetts
Printed in the United States of America
Book design and typography by Beverly Stiskin
Cover and illustrations by Richard Spencer

Table of Contents

Foreword 7
Introduction 9
1. *Mysticism and Science* 20
2. *The Path to Higher Consciousness* 32
3. *Self-realization and Psi* 47
4. *The Mind-Body Connection* 61
5. *The Chakras* 80
6. *Ki and the Chinese Meridian System* 99
7. *Actualizing Human Potential* 120

Afterword 139
Appendices
  A. Chakras and the Autonomic Nervous System 143
  B. Meridian Test Experiment 145

Footnotes 148

FOREWORD

DURING THE MIDDLE AGES, scholars required nothing more than the authority of Aristotle—"He himself said it"—as sufficient foundation for their precepts. Church authority was bowed to where necessary, but scholar-scientists capable of orating Aristotle chaired university departments from Padua to Prague.

The world, in this respect, is little changed. Today's science has its Einstein, Newton, and Planck. Few modern scientists become enthralled at clear demonstrations of their perceptual shortcomings—and these few characteristically become unemployed. But some among the few simply hurdle the economic and scholastic barriers imposed by conformism and proceed to blaze new trails into the forests of unknown but knowable things. Hiroshi Motoyama is such a man. Let us be precise lest the point be lost. British physicist, John Taylor, recently invited colleagues to witness a young lad bend metal objects inside a sealed container psychokinetically. A controlled experiment. Among said scientists some charged charlatanism, left abruptly, and preferred not to be troubled by any demonstration of forces quite unknown to physics. Professor J. Hasted of London, England, and Professor W. Franklin of Kent, Ohio, while separately working with psychic metal benders, have observed similar emotional denials among their "scientific" peers. One need not belabor the point. The major fault of today's science is this failure to accommodate scientific fact.

Scientists with unimpeachable credentials have demonstrated that regions can be perceived at a distance with non-visual mechanisms (R. Targ and H. Putoff); that human psychodynamics are related to the planetary correlates pertaining at the time of birth

(Gauquelin); that the mind survives death (E. Kubler-Ross and R. Moody), and that consciousness precedes life (I. Stevenson and D. Kelsey). One need not lengthen the list to convey the tattered state of academic science.

Mystics, to the contrary, have with clockwork regularity presented similar views all along. What tenet separates the Christ and the Buddha? Were Emanuel Swedenborg and Paramahansa Yogananda so different? Among living mystics are the teachings of Jack Schwartz and Swami Rama not one?

Science and the Evolution of Consciousness is not unique in depicting the hypothetical nadi/chakra physiology of the seer. Such may be found in the most ancient of Sanskrit texts. Rather, this important book is unique in elucidating how contemporary scientific observational tools may be employed to unravel the miraculous. No other work known to me both thoroughly encompasses a comprehension of the subtle chakra bodies and also presents new working tools whereby the facts of such subtle systems may be scientifically apprehended.

So, now we must wait. The drudgery of continued research must be undertaken. Observations must be compiled and conclusions arrived at by scientists in many countries and in various institutions. Such things take time.

But one can now announce with confidence that, should the Motoyama AMI device in scientific hands prove effective in revealing the existence of the chakra system, this discovery will be heralded throughout the world. Dr. Hiroshi Motoyama must then be ranked alongside Galileo and Leeuwenhoek, for future man will never again perceive himself in the same way.

E. STANTON MAXEY, M.D.
Florida Heart Institute

INTRODUCTION

Toward the Discrimination of Subtle Energies

WE MAY liken conventional scientific understanding of the universe to the visible tip of an iceberg. From our research studies over the past few centuries, we have come to know that exposed tip very well. However, most of Nature is still hidden from us and we know it not. History contains references to and speculation on many aspects of the hidden iceberg and recent research in the psychoenergetics area suggests some important features. From these studies, we appear to be dealing with new energy fields quite separate from those known to us via conventional science and upon which our present technological society is built. These same energies appear to function within our organism and strongly influence our behavior in the physical world. It is time we understood these energies that reach so close to the core of ourselves!

Today is the time for the maturing of these new studies and new ideas and, as we start mining, processing and extending them, we shall develop a new dimensionality to science every bit as concrete, as quantitative and as consistent as our conventional science. On this foundation will be built vast new technologies and new personal philosophies of man's relationship to each other, to the earth and to the universe at large.

As a step along the path, in this introduction I wish to outline some experimental data that violate our present World Picture and suggest some directions of needed change. The subsequent chapters of this fine book by Dr. Motoyama take us a good piece further down the path.

Cracks in Our World Picture

Up to the present, medicine, biology, and agriculture have largely viewed living organisms as operating via the following sequence of reactions:

$$\text{function} \leftrightarrows \text{structure} \leftrightarrows \text{chemistry} \qquad (1)$$

As a modification of eq. 1, there is some growing awareness of the interaction between chemical states and electromagnetic fields as the next component on the right in eq. 1.

Generally, flaws in the function area have been traced to structural defects in the system that arose out of certain chemical imbalances. The rectification procedure was usually via an adjustment of the chemical environment with more and more sophisticated chemical complexes being utilized to trigger the organism's defense and repair mechanisms. The dilemma that arises is that both the organism and the threatening invaders adapt to the new chemical complex becoming progressively less sensitive to it and so the escalation of potency must continue. One very deleterious aspect of this procedure is that the unnatural chemical content of the organism increases and begins to influence other levels of functioning of the organism than the one being corrected. The effect is particularly serious in the agricultural area where the method of application of the chemicals is via the soil so that a chemical equilibration develops between the plants and the soil, percolation of water through the system spreads the chemicals over a large area and the whole ecosystem begins to suffer from chemical pollution. Clearly, mankind must find a better way of understanding and dealing with flaws of function in living organisms. However, so long as he continues to view living organisms via eq. 1, man is stuck with his present methods.

In searching out alternative procedures for influencing the well-being of living organisms, one must first question the validity or completeness of eq. 1. Are there effective physical, as distinct from chemical, techniques for modifying organismic functioning? Are there potential techniques for doing likewise in the domain of what would presently be called nonphysical energies? Let us look at a number of observations that reveal the total inadequacy of eq. 1.

The most obvious discrepancy is the neglect of applied electromagnetic fields on the influence of muscles and organs in the human body. Everyone is familiar with the use of x-rays for tumor treatment and of diathermy for muscle relaxation, and we know that osteopathic physical manipulation techniques have had great success with improving human functioning for the past hundred years. In addition, from the field of neuropsychiatry, we have found that small electric currents between certain specific points in the brain give rise to the same behavioral changes that are observed with certain specific brain stimulating chemical intakes[1]. More recently, Becker has shown that small DC electric currents (1 $\mu\mu A/mm^2$ to 1000 $\mu\mu A/mm^2$) cause cell regeneration, tissue repair and fracture rehealing, whereas DC currents greater than 10,000 $\mu\mu A/mm^2$ cause cell degeneration[2]. Thus, the first step in modifying eq. 1 is

$$\text{function} \leftrightarrows \text{structure} \leftrightarrows \text{chemistry} \leftrightarrows \text{electromagnetic energy fields} \quad (2)$$

Going further, under hypnosis, the human body has been found to exhibit truly remarkable feats of strength and endurance attesting to a mind/structure link. In Aikido, Zen or Yoga disciplines, we see a conscious rather than a subconscious link between mind, structure and function. On another front, modern psychotherapy shows us that certain chemical treatments influence mental states and certain mental treatments influence chemical states[3]. Recent studies have indicated that human acupuncture points, of ancient oriental description, have different electrical characteristics than the surrounding skin[4]. It is very easy to show that significant changes occur in the skin electrical potential and impedance between waking, drowsy and sleep states as well as other hypnogogic states[5]. In fact, some Soviet investigators utilize concentration techniques to enhance the "effective" voltage difference between two acupuncture points from $\sim$ 50 mV to $\sim$ 500 mV as a training technique for developing psychokinetic abilities[6]. Finally, the recent development of biofeedback techniques[7,8] shows us that the directed use of mind can not only allow us to exercise control over a variety of autonomic body functions like skin temperature, pain, etc., but also to effect considerable repair of the vehicle. Certainly the experiments with Jack Schwartz thrusting needles through his body and mentally controlling the bleeding attests to this. From the foregoing, there is little doubt that "Mind" must be included on the right of our reaction equation; i.e.,

$$\text{function} \leftrightarrows \text{structure} \leftrightarrows \text{chemical} \leftrightarrows \text{electromagnetic energy fields} \leftrightarrows \text{mind} \qquad (3)$$

Now the going gets a little tougher; we have shown a link between what we have called mind forces and the other elements of our reaction equation in the human body. We have not demonstrated that these mind forces influence inanimate matter, nor have we demonstrated if there is an intermediate reaction component between mind and physical energy fields, nor have we demonstrated any unique and distinguishable characteristics of mind. In what follows, a partial satisfaction of these missing pieces to the puzzle will be indicated. Here, however, we must begin to go further afield to find published supporting data.

Careful studies of the enzyme Trypsin have shown that its activity can be altered by placing solutions of it between the poles of a strong magnetic field or between the palms of a "healer"[9]. The effect of the healer's hands was comparable to that found with fields in the range of 10^4 gauss. A more recent study[10] has shown not only the healer's ability to influence the growth of plants but also to influence cloud formation with the hands held (a) adjacent to a small cloud chamber and (b) adjacent to a visualization of the same chamber from 600 miles away. Reference to other healers, from Jesus to Arigo[11], continue to appear in our publications and one study[12] has even noted a change in the electrical impedance of specific acupuncture points on both the patient and the healer as a result of a laying-on-of-hands type of treatment.

Motoyama placed a person who showed psi-ability and a second, ordinary person, in separate rooms shielded by concrete walls lined with lead. The subject possessing psi-ability concentrated his mind on the other individual while Motoyama made measurements to determine if any changes in bodily functions occurred in the second (ordinary) person[13]. Motoyama found that remarkable changes in the pulse and respiratory rate of the ordinary person were evident during this concentration period. Since the two rooms were shielded against physical energy, he deduced that the psi-ability responsible for bringing about modifications was essentially nonphysical in nature.

From experiments directed towards man-plant communication, we gain more support for the interaction between nonphysical energies and physical energies. One of the most remarkable experiments is taking place in Findhorn, Scotland[14], where, in barren sandy soil and with a hostile climate, the small community has succeeded in producing dozens of vegetables, flowers, and trees in unexcelled size and beauty. Everything is based on the philosophy that views plants, soil, the natural forces of sun, rain and wind, plus people, as collective parts of the community of life. Via community meditation, talking to and expressing love and thanks to the plants themselves, and via invocation of the aid of elemental nature spirits, the people have created agricultural products which defy conventional explanation. Backster's initial work[15] and that of Vogel[16] indicated that an electrical response can be obtained from a plant consistent with the experimenter focusing his mind to project (a) the acts of damaging another plant, (b) the destruction of another life form or (c) a thought form of love, healing, etc. Backster postulated the existence of primary perception as a signal linkage between cellular plant life. However, Vogel postulated that the plants must be charged or sensitized by the experimenter's mind energies before being receptive to thoughts and emotions. Since a recent study[17] showed an inability to reproduce the Backster Effect without a plant sensitization step in the protocol, the Vogel approach may be the more relevant one. Finally, to see our connection to the earth, we note that the Schumann resonance[18] of the global capacitor between the earth and the ionosphere is in the 3 to 14 Hertz range which encompasses the α β, δ, and θ ranges of the dominant human brain wave patterns. Recent studies have revealed that 10 Hz electric fields control the human circadian rhythms and that human reaction times are strongly dependent on the specific ELF (extreme low frequency) field being propagated in the local earth environment[19]. In this series of experiments, we see a connectedness developing between living things at more subtle levels of energy than those associated with our physical chemistry.

Turning to the area of psychokinesis (movement of objects with the mind), we find an even greater confrontation for our present world picture. This author observed, first hand, psychokinetic demonstrations in the Soviet Union by two different people[20]. Swann performed two significant experiments in this area[21]: (1) the first, where he mentally altered the temperature of a thermistor, located at a fixed distance from his body, in a pattern consistent with a prescribed coding of hot, cool, neutral, etc.; and (2) in the second, he seemingly influenced the decay rate of a very sensitive magnetic field detector at Stanford University which was located behind vast metallic and concrete shielding. The psychokinetic experiments of Forwald[22], conducted over a 15–20 year period on small wooden cubes containing coatings of different materials of different thicknesses, showed an exponential dependence between psychokinetic force and coating thickness plus a proportionality to the neutron number of the coating material. Most recently, the demonstrations of Uri Geller have brought forth groups of children in England, Germany, Switzerland, Japan, etc., who possess similar and perhaps even greater abilities. Taylor[23] has recently described the experiments he has conducted with the young children in England. Perhaps most striking is their ability to mentally bend aluminum bars placed inside sealed transparent perspec tubes. The 1-foot long straight bars are seen to change into an S-shaped pattern filling the cross section of the 1" diameter perspec tubes without the end cap seals having been broken. Clearly, the mind forces can not only be used to influence living systems but inanimate ones as well.

Recent experiments on "remote viewing"[21,24] have shown that an individual can (a) perceive and accurately describe objects placed at a removed location from the perceiver and his line of sight, (b) be given the longitude and latitude coordinates of a location on the earth and accurately describe the terrain of that location even though it is thousands of miles away, and (c) tune in on a specific individual and view a remote locality through the individual's eyes. Interestingly enough, the sensitive individual sometimes perceives the scene before the target individual actually arrives on the scene. The reverse of this process, wherein one chooses a particular target, like a potential oil well, and seeks to know the specific earth coordinates, has also been accomplished. When one uses instrumental assistance, the technique falls in the category of "Radionics"[25,26] and the specific sub-category is called "map dowsing." Radionics is an instrumental form of radiesthesia (defined as sensitivity to radiations covering the whole field of radiations, physical and nonphysical, from any source either living or inert) and, as such, deals with the interaction of mind and matter and with the complete interrelationship of all things[27].

Conventional dowsing studies, wherein a type of wand is used and one walks over the ground being scanned, have shown[28] that the dowsing response is a muscular action connected via a sequence of biological processes to the cause which, in many cases, is a magnetic field gradient. One individual has shown himself to be sufficiently sensitive to respond to a 1 microamp current flowing in a wire, imbedded several feet in the ground, as he walks across the wire's location with the magnetic field at the wand level being about 10^{-10} gauss (earth field strength is 0.5 gauss). The seat of the sensing ability in the body appears to be the adrenal glands[28]. More recently, a number of U.S. osteopathic practitioners, investigating a subject, "kinesiology," have shown that specific body muscle tonus changes occur from specific minerals or chemicals merely held in the left hand or merely placed on the stomach (the ingredients may be in a glass container). This is a very similar result to that which this author has found using a type of dowsing wand called a biomechanical transducer[29].

Many studies have been carried out in the areas of telepathy and clairvoyance[24,30] attesting to their state of function in the human being and in animals. Karagulla[31] has utilized subjects with clairvoyant abilities to observe the "auric" fields around patients and thence describe their state of physical and mental health. These auric fields are perceived as patterns of light of different colors extending outwards from the body. With some subjects, the patterns can be viewed with the eyes closed or in total darkness.

Studies of precognitive awareness are also widespread and a precognition teaching machine has been developed[32] which allows one to train this ability into subjects, or, rather, to enhance the present operating level of this human capacity. Experiments being conducted with pyramids of specific shape suggest that they are useful for preserving cellular tissue and to enhance the healthy growth of plants, even in total darkness[30,33]. Combinations of specific shapes and specific materials in a unique fashion have been shown to produce devices called "psychotronic" generators[3] which can be charged with biological energy and perform an array of functions that confound conventional understanding. This author has directly observed such devices in operation and accepts that they most probably respond to a "New Physics."

The foregoing list could be made much longer; however, the point has hopefully been made that eqs. 1 and 2 are inadequate to account for a variety of new (and old) experiences that have occurred in the family of man. Acceptance of this work is difficult for those who have had no experiential awareness of these "other" energies of nature, and it is somewhat understandable that it is vehemently rejected in many quarters as violating our collective picture of the universe. Some of the natural criticism about sloppy experimental procedures is justified in specific cases because it is extremely difficult to develop a completely "clean" protocol for these experiments. In addition, one may be justified in quibbling about the quality of a particular experiment or about the veracity of a given experimenter; however, the body of experimental data of this type is so vast and growing so rapidly that it cannot be denied much longer. The evidence is becoming so strong that it definitely merits a wise person's reflection.

Incorporating the foregoing experimental results, we are led to a human reaction equation of the following form:

$$\text{function} \leftrightarrows \text{structure} \leftrightarrows \text{chemistry} \leftrightarrows$$
$$\text{positive space-time energies} \leftrightarrows \text{negative space-time energies} \leftrightarrows \quad (4)$$
$$\text{Mind} \leftrightarrows \text{Spirit} \leftrightarrows \text{Divine}$$

where the conventional space-time energies function in a reference frame which I have denoted positive space-time and the unconventional space-time energies function in the negative space-time frame.

From the foregoing, we can gain the idea that man is a multidimensional Being, functioning on many different levels of Nature simultaneously. He is mostly unaware of these levels of self and cannot grasp the visualization that he has an extended energy structure that interconnects and integrates his Beingness with seemingly separate localizations of Beingness. We are so chained to our view of reality as perceived by the five physical senses that we are unable to give credence to our true nature. We must come to realize the potential of man as expressed by eq. 4 and to recognize that, if man's energy structure is perturbed at any one of the indicated levels, ripples of effect flow out in all directions to produce corresponding perturbations at all other levels. However, the magnitude of the effect and the time of manifestation of the effect at another level will depend upon things like the intensity of the original signal, the conductivity of the medium of the original signal, and the degree of coherence of wave structures at the boundaries between levels.

A body structure at the physical level is stable and sustained by a type of chemical homeostasis or chemical pattern which, in turn, is maintained stable by a pattern of electrostatic potential and other specific positive space-time patterns of potential. The positive space-time patterns of energy are maintained stable by specific negative space-time patterns of potential which are themselves maintained stable by patterns at the mind level of the universe. At this point, it is useful to recall Wolf's law of bone structure which tells us that if a nonuniform stress is applied to a bone by the body for an extended period, the bone will grow new trabeculae (a type of bone girder) in the exact locations needed to support this nonuniform stress distribution. The process probably occurs by the physical strain field interacting with the electrostatic field of the system producing changes there, and these changes cause ions and molecules to be transported to specific locations agglomerating into specific tissues and structures that comprise the trabeculae. These changes probably all occur as a response to the strain energy contribution (pressure/volume factor) influencing the electrochemical potential, μ^j, of molecules in the physical tissues.

Carrying this idea further, mental field patterns can be thought to act like a stress influencing the field term of the magnetochemical potential of molecules, μ'^j, at the negative space-time level. In turn, via a special coupling, this pattern produces the required correlate at the positive space-time level in the μ^j, and thus in the structure of the physical body.

We must note that the removal of the body stress that created a certain pattern of trabeculae in a bone does not lead to the instantaneous dissolution of these trabeculae. Rather, they may disintegrate or dissolve very slowly (under the proper exercise) because of the molecular kinetics involved and they may maintain the body in a distorted shape for a very long time even though the initial physical cause is removed. The same situation occurs for physical structures generated by emotional or mental stress patterns. Further, since these unharmonious patterns scatter energy from the main flow stream at the various levels already discussed, the removal of the anchoring patterns at the two ends of the chain will release the intervening pattern links and more energy will be available for the organism's functioning.

All this leads quite naturally to a perspective on healing; i.e., that pathology can develop at a number of levels and that healing is needed at all of them to restore the system to a state of harmony. The initial pathology begins at the level of Mind and propagates effects to both the negative space-time and the positive space-time levels. We then perceive what we call disease or malfunction at these levels and try to remove the effects by a variety of healing techniques. The best healing mode is to help the individual remove the pathology at the *cause* level and bring about the correction by a return to "right thinking." The next best healing mode is to effect repair of the structure at the negative space-time level. The third best level of healing is that which medicine practices today, wherein one effects repair of the structure at the positive space-time level.

Since the energy structures at these different levels are coupled, repair at a lower level will still produce some feedback modification of energy structure at a higher level. However, if harmony is not restored at the higher level, then a force will continue to exist for pathological development in the energy structure at a lower level. Of course, this force is basically like a thermodynamic potential to produce change so that the effects may be manifested or materialized in very different forms, depending upon what alterations have already been made to the energy structures of the positive and negative space-time frames. The closest analogy to this can be found in the field of "phase equilibria" of materials. If you heat a complex alloy containing a number of chemical constituents to a high temperature so that it melts, then by cooling it again you produce a thermodynamic driving force for a phase change; i.e., to one of several possible solid forms. By making very slight but specific modifications to the chemistry or cooling rate or other variables in the process, it is possible to change the type of solid phase that initially develops and the crystalline form that results. However, so long as the potential exists, some solid phase will form.

What we can see developing here is a subjective science on an equal footing with present objective science. A common misconception is that the "scientific method" requires the experimenter to be coldly and distantly objective during the performance of an experiment. Instead it really only requires a complete description of the necessary and sufficient conditions of protocol for anyone to reproduce the experimental result at any physical location. If this requires setting a mental and an emotional field at certain strengths, then so be it. If these are to provide a positive, negative or neutral bias, then so be it. However, we must, of course, invent the measuring instruments with which to set and reproduce these field strengths. That is our problem today—we don't yet know enough to design the correct instruments; we only know enough to recognize a significant influence and to realize that it is not easily controlled. This recognition and its incorporation into the scientific method forms the basis for our "subjective science," which is of profound importance for understanding and guiding the transformation of humankind towards higher levels of integration.

There is a truly great need for reliable experimental devices for monitoring body energies on successively more and more subtle levels. Measurements with such devices will help to forge the bridge between our present chemical medicine and our future energy medicine. The studies discussed in the subsequent chapters of this book are a great step forward toward that goal!

William A. Tiller, Ph.D.
Stanford University

CHAPTER 1

Mysticism and Science

SCIENTISTS EXPLORE the diversity of life; mystics experience the unity. The aim is knowledge, and both groups expend a tremendous amount of effort to reach their goal. Each discipline demands highly sophisticated methods, years of training, and finely honed tools—requirements of the quest to understand the essential nature of reality.

In earlier civilizations, such as classical Greece or ancient China, mystics and scientists were neither separate nor antagonistic. But during the succeeding centuries, particularly in the West, the two groups diverged, to the point where they became mutually exclusive. Unity and diversity, spirit and matter, mind and body were split.

One of the more exciting aspects of the present age is that these traditions are beginning to reconverge. Advances in the natural sciences and the increasing accessibility of information about the mystic experience are forcing the two groups to take notice of each other again, with growing respect. Evidence is accumulating which indicates that the most advanced scientific and mystical characterizations of reality validate, rather than negate, each other. And the likelihood of ever-deepening cooperation between the two schools opens vast new frontiers of investigation, promising an enormous leap in humanity's understanding of itself and the reality of which it is a part.

Beginning with the Greek atomists, who formally insisted that spirit and matter are separate, scientific exploration has focused solely on the material world, the world perceived by the senses. The realm of spirit was left to philosophers and the church, which exerted widespread control over all aspects of culture throughout the

Middle Ages and held that knowledge of the physical dimension is unimportant. The growth of the natural sciences was therefore fairly stagnant until the Renaissance, when people began to struggle to throw off the decidedly repressive attitude fostered by papal dominance.

This shift saw the birth of modern science, a new science which took basic nourishment from the philosophy of René Descartes and the physics of Isaac Newton. Together, the theories of these two men convinced the majority that mind and matter are completely unrelated entities; they thus created the philosophy of materialism as we know it. Cartesian dualism asserts that since mind and matter are distinct (an assertion as yet unproven), it must be possible to dissect the world objectively; hence, the scientist's own effect on the situation under study can be disregarded. The underlying concept here is that matter is inherently real, that it exists independently of the consciousness observing it, and, furthermore, that nature can be dealt with as though it were a complicated but comprehensible machine. Isaac Newton relied on this axiom when developing his mechanics, which became the basis of classical physics.

This mechanistic approach to the universe dominated and directed western civilization until the beginning of this century, when it was undermined by the theories of relativity and quantum mechanics. The concept of materiality has effectively been destroyed for anyone who possesses the esoteric knowledge of relativistic and subatomic physics, but the far-reaching philosophical implications of this revolutionary realization are just beginning to touch the general public.

In the scientific view, the world is divided into subject and object. The object, all that can be perceived by the senses, is further fragmented into millions of related objects, parts of the whole. The task of natural science, accordingly, has traditionally been to "take apart" reality. The idea was that with enough effort humanity would be able to discover the basic "building blocks" of the universe and figure out how they fit together. Scientists felt that the workings of nature would be comprehended only if the essential constituents of matter could be determined, and their relationship to one another understood. The building block hypothesis ruled scientific endeavor for hundreds of years.

Here humanity was showing great faith in the simplistic notion that the most obvious tools provided by nature for understanding the world—the five senses and their corresponding consciousness—are true reflectors and interpreters of reality. The reality that our senses present to us is three-dimensional; in it time and space are absolute, and subject and object are perceived to be distinct entities.

The mechanistic world view, which depends on such sensory perception, served its masters very well: through it they accrued a staggering amount of intellectual knowledge and produced the technological revolution. Today it determines the way in which most of us construct our own personal realities and the way in which we live our lives. Yet this view has recently been invalidated by scientists themselves.

Ordinary, sensory perception does tell us that the world is made up of a collection of objects and events. Yet the substantiality of this information has been questioned by mystics of all cultures throughout history. The basic tenet of the mystical world view, regardless of cultural or doctrinal background, is that the multitude of things which we experience as distinct and real are but manifestations of the Absolute (undifferentiated reality)—that there is an underlying unity amid the seeming diversity of existence. It is only the individual, subjective mind—the consciousness informed by the senses—that fragmentizes the world.

Mystics further assert that the rational intellect, though it has a vital function in terms of biological survival, is inherently incapable of comprehending the unified nature of reality. It is therefore unable to deal with humanity's spiritual, psychological, and religious needs, all of which demand a deeper understanding of the true nature of existence. Fortunately, we are not limited to the rational intellect: other potential forms of perception/consciousness, now dormant in humanity, transcend the subject-object duality characterizing the type of consciousness that depends on the five senses (sensory consciousness). These other modes of consciousness must be awakened and developed if we are to experience reality and comprehend the meaning of existence.

Modern science is moving beyond a concept of reality bound by the dictates of ordinary perception, and coming closer to the view of the mystics. Einstein's theory of relativity and its correlates have shown that space and time are not absolute, but interconnected; they form a continuum now known as space-time. (Time and space, separately, are merely constructs of the human mind, relative to the position and velocity of the observer.) Relativity is now accepted as a working axiom of science, but our minds are still unable to experience the space-time continuum in normal consciousness, because of the limitations of our sensory conditioning.

In *The Tao of Physics*, Fritjof Capra suggests that modern physics has been forced, the more deeply it has probed seemingly dissimilar elements of nature, to admit the interdependence of all things. Concerning relativity, he says:

> Unification of entities which seem separate and irreconcilable is achieved in relativity theory by going from three to four dimensions. The four-dimensional world of relativistic physics is the world where force and matter are unified, where matter can appear as discontinuous particles or as a continuous field. In these cases, however, we can no longer visualize the unity very well. Physicists can "experience" the four-dimensional space-time world through the abstract mathematical formalism of their theories, but their visual imagination—like everybody else's—is limited to the three-dimensional world of the senses. Our language and thought patterns have evolved in this three-dimensional world and therefore we find it extremely hard to deal with the four-dimensional reality of relativistic physics.[1]

The limits of the rational intellect's ability to experience and understand nature have also been surpassed in the formulation of quantum theory. This theory was born from the experimental contradiction that matter sometimes acts like a collection of particles and sometimes like waves. Further investigation has shown that particles should not properly be designated as "things" at all, but as processes that cannot be defined apart from their interactions with other processes. There *are* no basic building blocks in the universe, only the continual interactions of related processes, which are, in turn, manifestations of the unified, holistic system which we call reality.

The role of scientists as "independent observers" is thus demolished. The way scientists set up experiments and the tools they use for measurement determine what they will be able to find. Werner Heisenberg said, "What we observe is not nature itself, but nature exposed to our method of questioning." Scientists do not work "in a vacuum"; they cannot blithely characterize what they experimentally discover in nature as being absolutely "real," for the way in which they are conditioned to think will determine to some degree what they are able to discover. Whatever they find is not necessarily inherent in nature itself but more a reflection of the way in which their minds categorize the observation. That is, whatever they are able to conceptualize with the rational mode of intellect is not the reality, but only an approximation, a formalized construct which they project onto reality.

Thus, scientists are not truly objective; they are subjectively linked to the processes of nature that they observe. The scientist is not an independent ego locked in the shell of a body; the subjective, sensory-dependent mind of the observer is very much related to what it is able to perceive.

This discovery has profound implications for the age-old problem of mind-versus-matter (spirit/body; nonmaterial/material), a subject traditionally treated by philosophers rather than physicists. But here also we see an impending unification.

The nature of mind's relationship to matter is a problem that has obsessed the human intellect throughout recorded history. Classical science, however, excluded mind as a viable topic of research because mind was not seen as an object that could be studied experimentally. A further obstacle was the fact that mind, assumed to be nonmaterial, was associated with spirit and therefore assigned to the sacrosanct dominion of the religious sphere. All branches of science have traditionally felt the need to dissociate themselves from anything remotely religious or mystical in order to be recognized as "scientific." Mind, likewise, could not be investigated until it had been divorced from spirit and subjected to this outdated ideal of scientific objectivity.

In western thought, religion came to be equated with faith and science with reason. The two standpoints were generally considered irreconcilable. A strong mystical thread does run throughout western civilization, but it has been obscured by religious and scientific dogma. Because mystical vision generally has a universalistic character exceeding the confines of narrow sectarianism, many Christian mystics were forced to limit the expression of their experience if they wished to avoid being charged with heresy.

Scientists interested in the transformation of spirit were also driven into the shadows, because they attempted to transcend the seeming limits of the physical world. Yet the large body of alchemical literature attests to their number. Newton himself considered alchemy a primary area of research; the rationally "acceptable" part of his work that was extracted and used as the basis of classical physics comprises only a small part of his thought.

In the nineteenth century psychology tried to bring the study of mind out into the light. To satisfy the masters of reason, the scope of investigation had to be drastically limited. So as not to infringe upon the dictates of faith, the larger questions of spirit and divinity had to be left closeted in the sacred halls of the Church. And the prejudice against the mystical tradition held by both religionists and scientists closed both groups to the valuable information contained within the esoteric traditions.

By the beginning of this century, materialistic tendencies were so deeply rooted that many psychologists, who were supposedly studying the "psyche," came to the conclusion that all mental functions are essentially neural processes—that, in essence, "mind" does not exist. Any inexplicable mental phenomenon that did not fit into this hypothesis was ignored. Fortunately, Freud, Jung and similarly oriented psychologists examined the mind with more than materialistic considerations and began to reveal strata of consciousness that lie beyond and below ordinary individual awareness. Parapsychologists have subsequently conducted research that indicates the existence of nonsensory forms of perception/consciousness which transcend the limitations of three-dimensional time and space. Contemporary transpersonal psychology and psychosynthetic models exist which attempt to lead consciousness into a recognition of its own deeper nature.

The real nature of consciousness and the way in which it functions still remain largely mysterious for scientists, perhaps because they objectify it just as they do other areas of investigation. As physics has shown, total objectification is no longer a realistic way in which to explore any phenomenon. And, as the mystical tradition has held all along, it is impossible to understand the true nature of consciousness by dividing it into pieces to be categorized. Mystics insist that, to gain true understanding, it is necessary somehow to go beyond the dualism of the rational intellect and to unify with the thing itself, to experience it.

Though science and psychology have begun inadvertently to validate concepts long propounded by western mystical tradition and eastern religion, there still has been no concerted attempt to test some of the basic empirical claims about human nature made by religion. Gopi Krishna makes this point in *The Biological Basis of Religion and Genius:*

> The most rational way to attest to the truth of religion, and to accept or reject its claims, would be to test these practices and disciplines after making a comparative study of all the systems in existence in different parts of the earth, and then to pronounce a verdict on the basis of the results achieved. But to this day it was never done by any group of scientists, dedicating their life to this research alone, in the same way that innumerable groups and societies are doing in respect to the still unexplained riddles of the physical world. The decision to ignore the claims of faith and to refuse its admission into the province of science was thus taken without trial, a most unscientific way of dealing with an obstinate phenomenon of this type.[2]

Actually, there has never before been a time when such research was feasible. But now, because of the technological explosion that has occurred in the twentieth century, research of this type can be done. Cultural and interdisciplinary boundaries are crumbling; ancient prejudices are slowly dissolving. Today, one human being—aided by modern education, libraries filled with a vast spectrum of translated texts, unlimited travel opportunities, and an intellectual attitude more open and tolerant than that of our predecessors—may delve deeply into a wide range of subjects. In my own life and work, I have tried to show that the way of the mystic and that of the scientist can be pursued simultaneously, that the knowledge of one can be fruitfully investigated by the other, that the realms of divinity and nondualistic consciousness can be explored with the tools of science.

My life has been imbued with the eastern mystical tradition since childhood. When I was six years old, I was adopted by a visionary who had undergone much mystical experience. Odaisama guided my spiritual growth for forty-five years. She exposed me to a wide range of esoteric teachings and initiated me into a series of rigorous disciplines, spanning the spectrum of Shinto, Buddhist, Taoist, and Hindu theory and practice. I was encouraged to study

Sanskrit, Chinese, and classical Japanese, that I might grasp the subtler meanings of the texts. During the same period, I also received a thorough western education, taking Ph.D. degrees in philosophy, medicine (physiology), and psychology.

The types of spiritual discipline I underwent were also taken from many traditions: they included both exercises aimed at purification of mind and body (fasts, retreats, yoga asanas, pranayama, etc.) and meditation practices directed toward the expansion of consciousness. The physiological and psychospiritual alterations produced by these disciplines led me to conclude that human beings do possess a potential far greater than that which is usually recognized. Further, I have reason to believe that the actualization of this potential not only represents the inherent direction of our evolution, but is vitally important to the preservation of the species.

I pursued such study and practices sporadically as a child, tapered off as an adolescent, practiced very intensely for seven years during my twenties, and for the past thirty years have practiced an hour or so a day, while serving as the priest of the Shinto sect's Tamamitsu ("Sphere of Light") Shrine, which Odaisama founded.

Odaisama saw no need for strict adherence to any one spiritual system, for her own experience had suggested to her that there are established principles in the evolution of being and consciousness which run as an underground river through the many divergent systems. She felt that human beings have not yet come to an objective, comprehensive formulation of these laws. One factor which has inhibited this formulation is the fact that universal principles tend to be obscured and confused by the wealth of emotional and cultural interpretation surrounding the mystical experience of any given individual. Another problem is that formulations are often misinterpreted within the confines of the organized religions that follow in the wake of the mystics who actually originate new teachings.

Odaisama believed in the unity of human experience and in a universal process of psychospiritual evolution. Thus, she felt that any practice that opens the gates to this process is valid; each practitioner may need a slightly different type of practice, but none is inherently better. She considered it imperative—particularly in view of the threat of global destruction—that the laws of spiritual evolution and the nature of reality be clarified for all who seek to know them.

Odaisama progressively realized the seriousness of the world situation in a series of visions (all recorded) which began in 1932 and ended with her death in 1974. The first vision revealed to her that war would break out seven years later (this was to be the Sino-Japanese War) and that a world war would follow. On March 7, 1946, she saw the world being divided into three poles, leading to a twenty-five year period of relative stability and prosperity. In 1971 she predicted that these three centers would gradually polarize into two, placing the world in a perilous position and perhaps leading to war and other cataclysmic events around the year 2000.

The visions, though apocalyptic, did suggest an antidote: Through the reunification of science and religion, humanity might be able to reach a level of understanding about the true nature of existence such that the impending crises could be avoided or at least ameliorated.

Such a task is a vast undertaking, and my own work represents only a small step. The early stages of my research were devoted to a metaphysical and ontological formulation of the content of religious and paranormal experience. Two books resulted from this effort, *Comparative Studies of Eastern and Western Mysticism* and *The World of Religious Experience* (in 1960 UNESCO designated the latter an outstanding work in the field of philosophy). But I came finally to realize that a philosophical approach does not have the power and immediacy needed to change people's lives. I decided to turn to the tools of science, in order to try to verify the existence and mechanism of nonphysical reality.

From 1962 to 1964, after studying psychology and physiology at Tokyo University, I did parapsychological research under the auspices of Dr. J. B. Rhine at Duke University. Upon returning to Japan, I began to do my own parapsychological work. However, the universities and academic circles looked down upon this type of research at the time, and I was discharged from the university where I was teaching. I encountered many other difficulties, but in the long run these adverse circumstances freed me to devote all my time and energy to my own research, unhampered by the petty sectarianism of academia. Most of the research described in this book was conducted at the Institute for Religion and Psychology in Tokyo, which was founded in 1962.

Two basic problems formed the nucleus of the research: (1) how to investigate the reputed existence of nonmaterial dimensions, and (2) how to elucidate the mechanism by which this supposed nonphysical realm functions in relation to physical reality. The main area of investigation has been the connection between body and mind. I used the scientific method to examine the ways in which eastern physiology (the subtle bodies, centers of consciousness, and energy systems referred to in Indian and Chinese medicine), as well as the western anatomical model, are related to the functioning of sensory and nonsensory consciousness.

Clearly, nonmaterial existence cannot be verified directly by scientific means, since the scientific method relies totally on the investigation of objects which can be cognized by the senses. Even subatomic physics, which is moving beyond the subject/object dualism, can only measure phenomena as they manifest in our sensory dimension. Particles, for example, can be detected by the tracks they leave in a bubble chamber, and the observation of these images, rather than the particles themselves, gives rise to present-day theories.

Total dependence on direct sensory observation locks one into the physical dimension. However, the scientific method, if skillfully used, can enable one to trace nonmaterial phenomena as they manifest on the material plane. Mystics have stated repeatedly that the body is the mirror of the mind, and that, in fact, all aspects of existence are similarly interrelated. ("As above, so below," is a generally accepted tenet of the mystic view.) Thus, we can hypothesize that any alteration in consciousness should produce corresponding psychological and physiological changes, changes that can be measured by the tools of science.

This hypothesis is grounded in the basic assumption that one can surpass the limits of individual consciousness, thereby attaining increasing awareness of other dimensions of reality. Individuals who claim to experience this breakthrough allegedly awaken levels of consciousness that are not dependent on the dualistic sensory apparatus and are thus broader than rational intellect. It seems probable that an individual who has attained this breakthrough would exhibit objectively verifiable paranormal awareness and physiological changes as a result of the experience.

My research therefore required a large group of subjects who had experienced alterations in consciousness primarily—but not exclusively—as a result of religious practice. In my role as priest I have come in contact with many such individuals, and as a scientist I have traveled extensively in order to find suitable subjects.

For purposes of comparison, I devised a tripartite classification. Group A includes individuals who claim to have attained a major transformation in consciousness. Group B consists of those who are in the process of attaining transformation. Group C is made up of those who do not feel they have had significant experiences of nonordinary reality; they constitute the control group.

Chapter 2 serves both as background and as a theoretical framework for the research; it relies on traditional texts to describe, briefly, how the breakthrough into nonsensory realms of consciousness occurs. A common manifestation of this breakthrough is the awakening of paranormal abilities, and these will be discussed in chapters 3 and 4, along with the parapsychological research I have done with the three groups. During the parapsychological work it became empirically evident to me, moreover, that the three groups had distinct physiological characteristics; these are discussed in chapter 4.

The neurophysiological framework, however, was not able to account fully for either the physiological differences among the groups or the paranormal abilities exhibited by members of groups A and B. To explain the results, I needed to turn to other models of human nature. The one that yielded the most helpful information was the *chakra-nadi* system developed over thousands of years by Indian mystics. This approach assumes an energy system in the body subtler than the one on which the physical system is dependent. The chakra-nadi system encompasses seven subtle energy centers known as chakras, which lie along the cerebro-spinal line, and a pervasive network of energy channels, the nadi.

The next step of my research was to investigate the existence of the chakras; to see whether any evidence of the chakras could be found using the scientific approach. The results of this research are described in chapter 5.

Another parallel system, one seemingly more directly connected with the physical system, is the meridian system delineated by Chinese acupuncture theory. Through years of meditation practice, I had become aware of a basic vital energy circulating throughout my body, and I came to suspect that this was the same system that forms the basis of oriental medicine. It became apparent in my research that this subtle system of energy is directly connected to both the chakras and the physical body; in fact, it may be the mediator between them.

I first endeavored to verify the existence of the meridians scientifically. The results are described in chapter 5. To elucidate the process by which the meridians and chakras function, in terms of consciousness and physiology, I invented two machines: the AMI machine (a diagnostic tool for measuring the condition of the meridians) and the "chakra machine," which is able to detect energy emitted from those regions of the body traditionally identified as the locations of chakras. Detailed comparative studies involving assessment of the three groups by these machines are now in progress. Some of the early results of this research are reported in chapters 6 and 7.

Chapter 7 also deals with some of the practical applications and philosophical implications of the work. I have begun to establish psychophysiological criteria to distinguish between pathological and paranormal states, for it is empirically evident that in the process of spiritual evolution the individual often goes through periods of physical and mental instability. This type of research may help to clarify the path consciousness takes as it develops, and this clarification should make the path safer and more accessible to those who would tread it.

Ultimately, I hope to aid in the creation of a new model of human nature. Accordingly, I intend to continue scientific investigation of acupuncture meridians, chakras, and vital energy, hoping to clarify the relationships between *mind—subtle body—physical body* and *paranormal abilities—chakras—vital energy*. Judging from the results thus far, I feel there is much promise in the direction my research is taking. My greatest hope is that we shall gain a deeper knowledge of who we are and how we are related to the universe, and that this knowledge will further the evolutionary progress of our collective consciousness.

CHAPTER 2

The Path to Higher Consciousness

THE COMMON notion that we are separate and distinct from the world around us arises from the way everyday consciousness experiences reality, through the five senses. People rarely question this mode of perception/consciousness; they assume that the senses act like windows on the external world, conveying faithfully what is "out there," and accept the notion that sensory perception is our only possible means of access to anything outside the internal "I" awareness. People tend to assume that the sensory mode of experiencing reality is a given and inviolable characteristic of the human condition, an absolute.

Mystics, on the other hand, assert that sensory perception/consciousness is only one possibility out of many—that it is not an absolute limit of the human condition, and that, in fact, its ability to apprehend reality is extremely circumscribed. Sensory consciousness may be the basic state connected to the physical plane into which human beings are born, but the mystics see it as being only the beginning of the evolutionary potential inherent in the human condition.

The senses are capable of only one thing: transmitting physical stimuli. The impulses that reach the brain through the network of sensory neurons are all the same. Those from the nose, the pain receptors, the eyes, are all identical. Whether we experience these as smelling, pain, or seeing depends on which part of the brain receives the impulse. The actual experiencing, therefore, happens totally inside the brain.

Evelyn Underhill[1] likens the conscious self to a person sitting "at the receiving end of a telegraph wire," and compares the sense receptors to the mechanism that is in touch with the external environment at the other end of the line and transmits its impressions as coded messages to the person on the receiving end. Internal consciousness does not interact with the outside directly, but relies entirely on the mediation of the senses.

Internal consciousness—sitting in the dark, as it were—is deluged by a continual stream of dots and dashes coming in over the wire. Its job is to censor out those it deems unimportant, and to create order and meaning among the rest. In other words, the primary function of this type of consciousness is to create a model of reality from the indirect, approximate information received. What we experience as reality is only the subsequently created model, a personal construction built by the mind. When we insist that we see a table, that the table "is," we are actually seeing an image in our own minds, minds that have in no way communicated directly with the table.

It follows that, inasmuch as consciousness depends on the senses for the information from which it builds its concept of reality, consciousness has separated itself from that reality. All that we can know for sure, from the dots and dashes, is that there is something beyond our imprisoned selves. To claim real knowledge about that something from the scanty information provided by the senses, say the mystics, is extremely presumptuous.

The model of reality constructed by the dualistic intellect is persistently mistaken for reality itself. Civilization was created and functions through our ability to agree on categories of meaning for specific packets of sensory input and to devise languages capable of expressing these distinctions. It is commonly agreed that what we perceive to be a rose *is* a rose. If someone were to say, "My, what a beautiful chrysanthemum!" when fourteen other people insist that it is not a chrysanthemum but a rose, we must assume that the odd individual's perception is faulty. This system of consensual reality may not be infallible (mass delusion is possible), but it is the universal code to which humans now subscribe. Misunderstanding occurs because it is easy for us to assume that mutually agreed upon, subjectively created representations of reality are the real thing; we forget that we are eternally separated from the "real thing" by the telegraph wire of the senses, and that the senses provide only limited and not always accurate information, and that we have created "reality" with our own minds.

We do, however, commonly acknowledge that the senses are

able to transmit only a selected portion of the seemingly infinite stimuli available. The eye, for example, responds only to a tiny slice of the electromagnetic spectrum. The same is true for all the senses, which are bound by the structural limits of our species. (Dogs, for instance, are able to hear sounds inaudible to humans; their sense of smell is more acute.) Still, an enormous number of stimuli fall within human range, and these bombard us continually. A major function of normal consciousness, consequently, is to sort this vast amount of data into stable and workable categories. It accomplishes this task primarily by relegating to the subconscious information that serves no immediate purpose and is not relevant for survival.

This task is so overwhelming that survival in fact demands that much of this analysis be taken over by subconscious, automatic function. Each of us possesses an individual unconscious which stores a huge amount of information and dictates much of our action. Having learned once that fire burns, we quickly become habituated to this notion and henceforth automatically refrain from sticking our hand into a flame. We do not need to repeat the reasoning process. In a similar fashion, we unconsciously tune out the constants in our environment and attend only to those aspects which change or are unexpected.

This type of division into conscious and sub- or unconscious function is paralleled in the nervous system, which is divided into two branches, the voluntary and the autonomic. The "unconscious" autonomic system is responsible for regulating and maintaining the complex physiological systems of the body. It enables the heart, for example, to function unattended by conscious awareness. Only when there is some change in normal functioning, some abnormality, does the unconscious thrust a message into the field of awareness (for example, "pain"), in order to make us take notice and act in some way to remedy the situation.

The great advantage of sensory consciousness lies in its miraculous ability to provide us with a stabilized model of reality—to create a relatively solid base of operations from which the individual can deal with an infinite number of physical stimuli. In short, sensory-dependent consciousness makes biological survival possible. The greatest disadvantage of this level of consciousness, on the other hand, is the fact that it gives rise to an "I" consciousness which, in Plato's words, is "imprisoned in the body like an oyster in his shell," inducing the belief that this mode of perception is the only one that exists—that the sensory "telegraph wire" is the only possible channel of communication between us and reality.

People often start out on the mystic path, the journey toward unification with nonsensory states of consciousness, the moment they begin to mistrust the validity of the way in which sensory consciousness perceives reality. Suddenly the heretofore "real" world appears to be nothing but an illusion. One begins to feel constricted and trapped by the limitations imposed by ordinary awareness. This dissatisfaction gives rise to a desire to become free from these dictates, to participate in reality more directly, to seek union with something as yet unknown which is greater than the individual self locked inside its "shell."

An ample body of empirical evidence indicates that it is definitely possible to break through the shell of individual consciousness—that many other forms of consciousness do exist and are universally available. After detailed research into the nature of consciousness, William James concluded that "our normal waking consciousness, rational consciousness as we call it, is but one special type of consciousness, whilst all about it, parted from it by the filmiest of screens, there lie potential forms of consciousness entirely different."[2] Mystics tend to characterize the various forms of consciousness as stages in the evolution of existence; they claim that the stages become subsequently higher or broader and that their progressive realization represents the actualization of human potential. Whereas ordinary consciousness is subjective, fragmentary, and dualistic, the nonsensory states are said to be increasingly less so; they move steadily toward universality, and thus a more immediate understanding of reality.

If it is true that we ordinarily experience reality through one type of consciousness and that other modes are possible, how do we attain them? By what process can an individual break through into other dimensions of awareness?

Countless techniques have been employed in all cultures throughout history to conquer the sensory-dependent self and actualize other states which more objectively reflect reality. A study of these techniques, as well as independent descriptions of spiritual experience, suggests that there exists a coherent process of mental development, a given manner in which mind evolves, comparable to the development of the human body, which can be willfully accelerated. One of the most prominent features of traditional spiritual paths is this emphasis on process and technique, rather than result. The reason for this approach is that it is universally agreed that the contents of nonsensory states of consciousness cannot be expressed in the languages devised by the dualistic intellect (though often those who experience them attempt to give at least an approximate description). The fact that it is difficult to talk about these states does not disprove their existence; it only points out the inability of sensory-dependent intellect to apprehend and deal with experience beyond its ken. In fact, the identical themes that recur in the characterizations of mystical experience, regardless of age or culture, and the discernible similarities in the numerous techniques used to produce them are clear indications that something important and universal does happen when consciousness is pushed beyond its customary limitations.

There are literally thousands of texts and teachings—by Christian mystics, Kabbalists, Buddhists, Sufis, Taoists, Tibetans—to help guide one through the psychophysiological process of realizing nonsensory states of consciousness. To help explain the type of claims I have attempted to investigate in my research, I shall describe one of these traditional approaches as background. From the diverse accounts available, I have chosen one characterization of the process which seems uniquely comprehensive, Patanjali's *Yoga Sutras*,[3] compiled in India sometime between the second century B.C. and the fifth century A.D.

The Indo-Aryans have been preoccupied with trying to understand the nature and development of consciousness for many centuries. Thus, two thousand years ago, when Patanjali was born, there already existed in India a multitude of sects and teachings based on the experience of countless mystics and yogis. Patanjali's major contribution was to cull the basic facts of the psychophysiological process of transformation from all this information and to codify these into eight stages, known as the eight limbs, which he set down in the *Yoga Sutras*. Meditation is divided here into steps in order to examine the overall process; these steps do not occur as discrete, nonoverlapping segments but represent convenient distinctions in an otherwise unified experience. Pantanjali managed to discard a great deal of secondary information and subjective interpretation surrounding the process, producing a description that is clear, concise, and surprisingly free of dogma. One might even say it verges on the "scientific."

As codified by Patanjali, the eight stages are:
1. Rules of abstention (*yama*)
2. Moral precepts (*niyama*)
3. Physical postures (*asana*)
4. Regulation of respiration (*pranayama*)
5. "Emancipation of sensory activity from the domination of exterior objects" (*pratyahara*)
6. Concentration (*dharana*)
7. Meditation (*dhyana*)
8. Union (*samadhi*)

The first two stages help to balance the mind and to teach self-control in preparation for the work. The next three stages are necessary for the development of concentration and the liberation of the mind from its normal activities, those which depend on physiological energy. The last three are the tools with which consciousness can crack through the shell of individuality and unify with or expand into other forms of consciousness. Let us examine each of these stages in greater detail.

Yama and Niyama

The first two stages, abstention and precepts, teach self-control; they begin to disentangle one from the domination of egocentric desires. The two function together to help the seeker gain control of negative thought and action. Both are found in some form in all religious systems. (The form most widely accepted in the West is the Ten Commandments.)

From the psychological viewpoint, it can be said that the ability to refrain from destructive actions and the conscious attempt to lead a good life are necessary for the maintenance of mental health. A negative act does not simply disappear: it gives rise to various complexes in the subconscious that consequently pose a threat to the conscious level, and may eventually disturb the balance of the mind and the workings of the mind/body relationship, which can manifest as disease, neurosis, or other related problems. The negativity may spread outward, influencing personal relationships and disturbing the peace of society in general.

Indeed, the striking number of individuals suffering from mental and emotional problems in contemporary society may be attributable to the fact that, with our relatively recent rejection of religion and the attendant morality, we have simultaneously lost the concept of any higher existence. Vainly, we have elevated ourselves to a position of eminence where we assume the right to affect the universe as we please. Without the awareness of a divine entity or some higher form of consciousness to which we are ultimately connected, we have little motivation to maintain a high standard of morality. And it is becoming increasingly obvious that neither internal nor external harmony can be maintained without some degree of self-control. This is true in daily life, and even more so in the case of someone who is trying to go beyond the ordinary human conditions of dependence on sensory consciousness. Without a basic level of control and stability no one can free the energy necessary to begin the process, or maintain the degree of effort and responsibility required for the awakening of other states of awareness.

Asana

The third stage of Patanjali's description, asana, consists of the practice of certain physical postures. Often these are maintained for a long period of time. When the body is thus concentrated into a single position, the mind is released from concern with physical movement, and attention can be turned to the field of consciousness. An important aspect of these postures is the fact that they also function to maintain a straight spine.

Let us consider two representative postures, those most widely used for meditation in the East, the full and the half-lotus. If done correctly, these postures allow one to hold the spinal column as straight as possible, to release all unnecessary tension from the neck, shoulder, lower back, and arms, and to maintain—without strain—the center of gravity in a spot about two inches below the navel.

Psychologically, anxiety locked in the body in the form of physical tension is a waste of energy and distracts from one's concentration. The meditation postures help to release tension. Muscle tensions in the lotus posture have been found to be lower than in any other body position except that of lying down. When all the organs of the body are in their proper places and thus no effort is needed to support them, nerve impulses, and blood, can flow freely. And once active consciousness is freed from having to control any physical processes, true relaxation is possible.

Physiologically, the importance of a straight spine is being increasingly acknowledged. Clinical examination of patients with chronic illnesses reveal a marked number of dislocated vertebrae. For example, patients whose third and fourth thoracic vertebrae are displaced often exhibit heart trouble or asthma, whereas those with abnormalities in the fifth to the ninth vertebrae show problems in the digestive system. One explanation may be that because the internal organs are each controlled by nerves emanating from a specific vertebra, a dislocation in that vertebra will put pressure on the nerve and thus interfere with the function of the corresponding organ. Conversely, when an acute illness appears in the internal organs and becomes chronic, an abnormality usually becomes evident somewhere in the spinal column.

This pattern is seen not only in physical maladies, but also in those of a psychological nature. Patients with nervous or mental disorders often display dislocation in the seventh cervical and first thoracic vertebrae. It is thought that these dislocations reduce the amount of blood able to reach a specific part of the brain. The ensuing deficiency may serve as a catalyst for abnormalities such as manic depression or schizophrenia. A singular benefit of meditation postures, therefore, is that they act to straighten the spine and thus can help to prevent or cure related maladies.

Pranayama

As the body is stilled through the practice of holding one set posture for a long time, so is the breath quieted through the practice of regulation. A yoga teaching states that if the mind is moving, so are the heart and respiration. When we are angry, our breath quickens; when we sleep, our breath slows down. By consciously slowing down the breath and making it rhythmic so that consciousness is not disturbed by it, we can achieve a corresponding tranquillity. The major physiological objective of the fourth stage, breath regulation, is accordingly to slow down and control respiration. This practice is usually accompanied by the sustaining of physical postures.

A representative method of pranayama is to inhale slowly to a count of four, hold the breath in the area of the abdomen for a count of sixteen, and exhale to a count of four. This method, like the majority of the many types of pranayama, teaches one to breathe from the abdomen. The practice of abdominal breathing has many beneficial effects which can be explained in terms of western medicine.

The human body has several nerve plexuses, which work as relay stations for the transmission of signals from the brain to various parts of the body. They may be likened to the marshaling yard of a railway station or to the central office of the telephone system. All the nerves controlling the functions of the organs in one's belly spring from the solar plexus, located just behind the stomach and below the diaphragm; it is such an important part of the body that it is also called the "abdominal brain." When one does abdominal breathing, the pressure in the abdomen rises, stimulating the diaphragm and the solar plexus. As a result, the functioning of the related organs is improved.

This improved activity is also true in terms of blood flow. The human body has two kinds of blood vessels, arteries and veins. Arteries actively convey oxygenated blood away from the heart; veins carry deoxygenated blood back. Arteries are controlled by nerves. Veins, however, have valves which regulate the flow of blood through them—they are not controlled by nerves but by muscle action. Therefore, the flow of blood in the veins is often weak or sluggish, especially in those who do little exercise. This sluggishness often leads to undue strain on the heart, because it is compelled to contract with abnormal force to compensate for the inactivity of the skeletal muscles.

Abdominal breathing requires that the skeletal muscles move more than usual; this affects the veins so that the blood is recycled more energetically and the burden on the heart is lessened. Subjects who practice sitting with an erect spine and breathing abdominally show an increase in the sharp incisions of their plethysmograph (the data provided by a machine that measures blood flow), indicating that the blood is thrust powerfully from the heart and that the overall blood circulation is healthy. Some doctors now recommend that patients with angina pectoris begin abdominal breathing when they feel an attack coming on.

Many different systems have used breath regulation as a technique in the attempt to alter consciousness. The Taoists, for example, teach a method called *tenporin*, conceived as a circulation of light. The disciple concentrates either on drawing the breath from the coccyx to the head and back to the coccyx (the *shoshuten*, or "lesser circle") or from the soles of the feet to the head and back down again (the *daishuten*, or "greater circle"). Various Christian meditators also mention an equivalent of pranayama practice. For example, in *Method of Holy Prayer and Attention*, Simeon, the New Theologian, (circa 1250 A.D.) instructs the monk: "Then seat yourself in a quiet cell, apart in a corner, and apply yourself to doing as I shall say. Close the door, raise your mind above any vain or transitory object. Then, pressing your beard against your chest, direct the eye of the body and with it all your mind upon the center of your belly—that is, upon your navel—compress the inspiration of air passing through the nose so that you do not breathe easily, and mentally examine the interior of your entrails in search of the place of the heart, where all the powers of the soul delight to linger."[4] Pressing the chin into the chest is a common yogic device to aid in the retention of breath.

Correct living through adherence to codes of morality, physical and mental stabilization through posture and breath regulation—these are accomplished through the practice of the first four stages. The next stage, withdrawal of the senses, continues the process of quieting the mind and liberating it from ordinary activity.

Pratyahara

The Indian tradition often likens sensory intellect to a wild monkey. Just as the beast swings in a constant frenzy from tree to tree, screaming all the while, the ordinary mind frantically attends to the inrush of sensory stimuli from outside and the uprush of memories, emotions, fantasies, from the unconscious. We "lose our minds" so easily—they wander ceaselessly amid external objects, fantasies, relics from the past. An important step in the attempt to realize other levels of consciousness, therefore, is to integrate fragmented awareness.

In regard to the stilling of the mind necessary for the growth of awareness, Meister Eckhart has said: "The soul, with all its powers, has divided and scattered itself in outward things, each according to its functions: the power of sight in the eye, the power of hearing in the ear, the power of taste in the tongue, and thus they are less able to work inwardly, for every power which is divided is imperfect. So the soul, if she would work inwardly, must call home all her powers and collect them from all divided things to one inward work."[5]

The ways devised to "call home" the soul's powers vary from system to system, but all require placing oneself in an environment and a psychophysiological state that make one less vulnerable to the intrusion of sensory stimuli. Traditionally, devoted seekers have isolated themselves as much as possible from the mundane world by retreating into caves and monasteries, by journeying to mountaintops and deserts. But such extreme measures are not necessary. One need only refrain from reacting to the stimuli of the outside world by allowing fewer data to come in over the telegraph wire and by inhibiting the mental activity of censoring the information still getting through.

Many effects of this sensory withdrawal can be detected and measured physiologically. For example, one of the common pratyahara techniques is to close the eyes, completely or partially. Electroencephalographs (machines which trace the changes in electrical potential produced by the brain), or EEG's, indicate a significant change in the brain's activity even when the eyes are simply opened, then closed. Alpha waves appear more frequently when the eyes are closed; beta waves appear as soon as the eyes are opened. Alpha waves are thought to correspond to a quiet mental state, beta to mental activity. Simple light stimulation will also effect many changes in the mind and body. A change caused by the stimulation of the retina by light sends electrical impulses to the occipital area of the brain, which governs visual perception. The impulses race on to the memory center in the temporal lobe, then on to the frontal lobes, where are stored the categories that enable us to distinguish, for example, whether a color is red or blue. Next, the brain relays nerve impulses to the body, causing many subtle and complex reactions, such as an increase in the secretion of glycogen (a substance produced in the liver which is necessary for muscle movement). In short, when light comes into the eyes, the brain and the whole body begin to respond automatically. Thus, by allowing in less light and keeping the source of stimulation constant, one can temporarily deactivate the functioning of the stimulation/response mechanism.

These few examples suggest how mind and body may be quieted through the practice of pratyahara. The activities of ordinary consciousness are suspended; the mind, which has been "lost" somewhere in the outside world, is regained. Now the work of real concentration may begin.

Dharana and Dhyana

Although the existence of the individual unconscious was not formally postulated by western thinkers until the mid—nineteenth century, it has been clearly acknowledged by the philosophical and spiritual traditions of the East for thousands of years. The experience of the eastern thinkers taught them long agó that ordinary, sensory-dependent mind has two parts—one we are actively aware of and one we are not. The individual unconscious is the storehouse of all egoistic desires—"egoistic" in the sense that these desires are a product of the illusion that self is separate from other (an illusion born from the sensory modality of perception).

Not only did the eastern (specifically, Hindu and Buddhist) traditions recognize the existence of the individual unconscious, they also realized its enormous complexity, depth, and power—its power to influence our thoughts and actions in a way that maintains the illusion of self as distinct from other, thereby preventing unification with nonsensory states. The individual unconscious is seen to be the greatest obstacle one must overcome in breaking through the sensorily created shell that limits us from participating more directly in reality. The eastern disciplines employ the tools of concentration and meditation to overcome this obstacle, to undermine the control that the individual unconscious exerts over us.

Concentration is the act of riveting attention on one fixed source of stimulation. Patanjali defines it as "fixation of thought on a single point." The mind and body normally exhibit varying degrees of activity, depending on the amount and type of stimuli received, but this activity is continual. It prevents unconscious elements from taking over the field of awareness. When one chooses a single point of concentration and holds on to it, awareness is forced to stop running around.

Directing all one's attention to one point produces a temporary simplification of the operations of the brain and corresponding sensory consciousness. Awareness is thereby released from its ordinary occupation, that of attending to a multitude of stimuli. Because this task acts as a barrier to the direct intrusion of the unconscious, once the field of awareness becomes quiet, the energy-packed contents of the unconscious spontaneously begin to spill over into the other territory. We see this happen whenever the normal function of awareness is suspended. (When, for example, someone is drunk or hypnotized, emotionally laden elements usually burst forth.)

The purpose of undergoing the mystic discipline, however, is not merely to understand the unconscious or to release its repressed elements, but to go beyond it, into totally other nonsensory forms of consciousness. These other forms of consciousness are usually lumped together in one category. Jung called them the "collective unconscious." Eastern religions refer to them as "transcendent" or "cosmic consciousness." Basically, these are the states of consciousness which are not locked into the limitations of the material plane, in which the distinction between subject and object is not rigid, in which the individual disappears into a unified relationship with all being. The importance of releasing the repressed energy of the individual unconscious through meditation is to free oneself from the domination of the repressed elements, so that one may unify with higher, nonsensory states.

The factor that enables one to become conscious of the unconscious is called "the observer" or "witness consciousness": the ability to watch the flow coming out of the individual unconscious without becoming involved in the drama—the ability to dissociate oneself. The observer is a spontaneous product of the practice of concentration, and the meditator is admonished repeatedly to strengthen the observer by paying attention to the object of concentration and letting go of whatever comes into his or her mind. As the observer becomes stronger, through continued practice, it becomes increasingly able to watch silently as various contents stream through the quieted field of awareness.

In the beginning stages of any concentration practice, the observer has not yet gained the power to withstand the deluge coming from the unconscious; it soon disappears as the field of awareness is overrun with memories, thoughts, feelings. When meditators remember their task, they again begin to redirect attention to the object of concentration, letting the subconscious content go right past them, paying no attention to it. This process is continually repeated, until the concentration span can be controlled.

As the subconscious contents continue to be released, they lose their energy—just as an uncapped carbonated beverage becomes flat. A state is finally reached in which ideas and memories rarely drift into the field of awareness. Then the mind becomes truly quiet. Such total silence can only be maintained for very short periods at first (a second or two), but the duration lengthens with practice. This silence is accompanied by a feeling of deep peace and indicates that the practitioner has entered the initial stage of meditation, a spontaneous psychic continuation of concentration requiring no new "technique." A common physiological manifestation of this stage is that the respiration rate, normally 16 breaths per minute, will automatically decrease, usually to around 10, but sometimes to as low as 2 or 3. Buddhists call this state *munenmuso*—"no thinking, no thought." Saint Theresa called it the "Orison of Quiet" and places it, in the exact same manner as Patanjali, after "Recollection" (voluntary concentration) and before the "Orison of Union" (samadhi).

Patanjali defines meditation (dhyana) as "a current of unified thought." A later commentator, Vyasa, enlarges this definition slightly, saying that meditation is a "continuum of mental effort to assimilate the object of meditation, free from any other effort to assimilate other objects." The attainment of the stage of meditation brings with it a change in the subject's relationship to the object, and a merging of the two begins to take place.

Samadhi

This stage is difficult to conceptualize, but we may elucidate it by going back to the metaphor of the telegraph wire. The consciousness, which was so busy attending to outside stimuli and being dominated by the unconscious, is now concentrated and still. It attempts, through continual active awareness, to communicate directly with what is outside itself, rather than depending on the senses for information. At the stage of meditation, the realization occurs that the object can be cognized directly, without the aid of the senses and mental categories, but there is still a lingering sense of the "I," of separation between subject and object. It is as if the shell surrounding the individual, the shell created by the sensory mode of perception, had become punctured with holes, and through these holes the subject is able to merge with the object and experience it directly. The longer this state is maintained, the larger the holes become, until the shell eventually dissolves and total unification between subjective consciousness and something greater than itself takes place. Patanjali enumerates a number of gradations of samadhi, which are all states of union but in which subtle illusions of self as distinct from other are still maintained. The final state is called *asamprajnata* samadhi, of which Mircea Eliade says: "The Yogin who attains to this samadhi realizes a dream that has obsessed the human spirit from the beginning of its history—to coincide with the All, to recover Unity, to re-establish the initial non-duality, to abolish time and creation (i.e., the multiplicity and heterogeneity of the cosmos), in particular, to abolish the twofold division of the real into object-subject."[6]

Though there appear to be many states and levels of consciousness beyond the physical, I am less concerned with providing a description of the states themselves than with examining the initial supposition that it is indeed possible to break through the individual shell into other states of consciousness. I plan to add more concrete information to the simplified, theoretical system just described, as well as to note some of the objective, measurable changes that occur in mind and body during the process. I shall begin by examining the basic hypothesis that if one is able to break through the limitations of dependence on sensory consciousness, nonsensory forms of perception and modes of interacting with the external environment will naturally follow. Let us now turn our attention to such "nonsensory" phenomena.

CHAPTER 3

Self-realization and Psi

MOST PEOPLE have had at least one intuitive experience for which no rational explanation can be found. One may have a piercing thought of someone one hasn't seen in a long time, and the following instant that person calls or turns up on the street. One may suddenly have a vague, inexplicable feeling of apprehension—then find out later that, say, the bus one decided not to take has crashed.[1] People often sense something is wrong when someone close to them has died; often parents know when a child they are separated from becomes ill or is in trouble.

Most people acknowledge this type of experience, but such instances are so rare that they are accepted without much thought and the phenomenon generally is not considered important. Some people, however, experience so many premonitions of such accuracy that they would appear to possess extraordinary powers of mind. Depending on the scope and intensity of their abilities, these individuals are called witches, psychics, prophets, or saints.

People who cling tightly to the sensory-dependent mode of consciousness—the majority—often react to these unusual individuals with fear or mistrust. Consequently, the meaning and mechanism of experiences that seem to defy the assumed limitations of the human mind have not yet been questioned or dealt with seriously by many branches of learning. Mystics, however, are well acquainted with these abilities and in fact view them as concomitant to the awakening of nonsensory states of consciousness. In ordinary life, such manifestations of nonordinary perception may seem haphazard and uncontrollable, but mystical tradition asserts that they are a normal function of nonsensory states and, as such, follow certain universal laws. Once one understands these laws, the range of possible manifestation far exceeds one's image of "psychic power." Such super-power is available to the spiritually mature. In this chapter, I shall summarize the mystic characterization of nonsensory abilities, state why I find them a valuable area of research, and give an overview of the contemporary parapsychological investigation being done in this area.

Included in the vast world of "psychic," or "paranormal," or "psi" phenomena are the abilities to see, hear, smell, taste, and touch objects without using the corresponding sense organs. Consciousness need no longer depend on, for instance, the physical mechanism of the eye to see; one can see beyond the dualistic sensory limitation of ordinary vision.

The more common varieties of nonsensory experience are the following abilities: to see things regardless of their spatial relation to the cognizer (clairvoyance); to receive psychological events directly from the mind of another person (telepathy); to know things that took place in the past (retrocognition) or will take place in the future (precognition); to effect behavioral changes in matter or mind without using sensory means (psychokinesis); to transport matter over distance or to dematerialize and rematerialize it without using physical means (teleportation); to heal the sick, injured, or insane without having recourse to orthodox healing methods (faith healing, psychic surgery). These abilities fall naturally into two categories: nonordinary perception and nonordinary action.

Although paranormal abilities may be developed systematically through discipline, one need not undergo such discipline to experience them. The mystics hold that any aspect of consciousness available to one person is ultimately available to any other. The various forms of perception and consciousness are there, to be experienced when the proper conditions are created. Mystic discipline is an ordered, tested method to achieve them, but they may occasionally occur spontaneously in anyone.

Empirical evidence has led me to believe that when the normal functioning of consciousness is altered or suspended for any reason (for example, through hysteria, shock, sleep, drugs, alcohol, or hypnosis) information from more unified levels of consciousness may occasionally break through. In the vast majority of such cases, however, the experience is not one of nonsensory consciousness, but merely a hallucination originating in the subjective, individual unconscious. Criteria are needed which can distinguish between objective psychic experiences and subjective fantasy in order that the objective experiences can be identified as such and then studied. Before discussing the specifics of this type of criteria, let us briefly explore some of the attitudes that the mystics take toward paranormal faculties and the explanations they give for the occurrence of nonordinary perception and action.

The traditional mystical attitudes toward these paranormal abilities are paradoxical. On the one hand, psychic abilities are recognized, classified, and seen to be objective indications that one has attained a corresponding level of consciousness. (The mystics liken the faculties to those with which we are endowed at birth: When babies are born into the physical world, they are equipped with eyes to see, hands with which to touch, etc. This is considered normal and natural. In the same way, say the mystics, when one breaks through physically dominated consciousness into other forms, one is spontaneously endowed with new faculties.) But, just as one must give up dependence on physical, sensory–created consciousness in order to realize nonsensory states, one must not become enamored of and controlled by the psychic powers, the mystics warn, or one will grow no further. The aim of evolution is liberation from the dualistic mind, not to separate oneself further from "the other" by building grandiose notions of one's own "powers."

It is easy to imagine how people might become fascinated by the fact that they possess extraordinary abilities, and how this self-importance might lead to abuse of the ability. Accordingly, mystical systems stress nonattachment to these powers, caution against their use, and state emphatically that the abilities are a by-product of evolution, *not the goal*. They specify that these abilities are not to be used for self-enhancement; in the end, a self-aggrandizing attitude would defeat the basic objective, that of experiencing the unified nature of reality. It is to help prevent the emergence of such an attitude that abstention and morality are found in the first and second place of Patanjali's system and figure in all religious systems: they help the individual to control selfish desires. The danger of self-aggrandizement is also a primary reason why people who possess truly developed and controlled abilities so rarely display them openly—why the entire "esoteric" tradition developed.

Patanjali devotes book 3 of his *Yoga Sutras* to enumerating and discussing psychic powers (*siddhi* in Sanskrit). He elucidates abilities which the yogi may attain. Among these are the abilities to know what is in other people's minds, and to understand the communications of animals. The explanation he gives for these phenomena centers on the fact that the yogi is able to unify directly with an object which is the subject of meditation—another person's mind or an animal's brain—and thus can gain immediate knowledge of every aspect of that object. By breaking down the barrier of sensory-dependent intellect, one may know the thing itself, rather than just a sensory-created approximation.

The awakening of nonordinary abilities is not uncommon among those practicing strict meditative discipline. I myself have had a great deal of direct experience with the paranormal, as have many people I have come in contact with over the years, including a large proportion of my research subjects.

One example, among a multitude of cases that have come to my attention in the past fifteen years, was reported by Sri M. Satyanarayana, an Indian yogi who is a devout practitioner of kundalini yoga. In a paper prepared for the fifth annual convention of the International Association of Religion and Parapsychology, held in Tokyo on November 14, 1976, he stated: "On 1-1-63 the Divine Power began to function in me controlling every movement . . . During these days some mystic powers manifested—I could understand the language of insects and animals and could also direct them to do my bidding—vocal music with which I was not conversant began to flow—foreign languages, when heard on the radio, were understood—clairvoyance and clairaudience came under control . . ." But the attainment of these powers was not the aim of his practice. It only happens, he pointed out, that success in kundalini yoga is indicated by the ability to control "the Universal Power"—"*whether one uses it or not.*" The yogi added that he actualized the powers "for a short time to know for certain that they manifested in me. Then I consciously refrained from exhibiting them as I am convinced that they are detrimental to further progress."[2] This same view is found in the *Yoga Sutras*, and all sincere seekers seem to adhere to this principle.

Unfortunately, however, there has been much confusion over the meaning and proper usage of psychic powers in many mystical traditions, leading to a general loss of perspective. Certain traditions, in the attempt to dissuade the adept from desiring nonordinary abilities as ends in themselves, have gone a bit overboard, either ignoring their existence altogether or exaggerating the possible dangers involved. As the by-product of the evolution of consciousness, paranormal abilities need to be dealt with in a more objective, matter-of-fact manner—particularly today, when so many people are experimenting with traditional meditative practices, often without the benefit of a competent guide. If the value and mechanism of paranormal abilities were better understood, people would have a more stable framework in which to view their own experiences of nonsensory perception and reality.

Problematic approaches to paranormal abilities abound. Two examples are the Zen Buddhist and Christian traditions.

During Zen meditation, as in any other kind, a lot of repressed content is unleashed from the unconscious; this content may manifest in awareness as visions and hallucinations. The manifestations are sensorily created; they are subjective. But as one opens to the nonsensory realms of being, information can enter the field of awareness which is not born of subjective experience—it is an objective experience of another plane of being. In Zen these two types of phenomena are lumped together under the term *makkyo*, which translates roughly as "manifestations of the demon realms." Most Zen masters teach their students to ignore makkyo—indeed, to avoid them at all costs. It may be that Zen simply overemphasizes the necessity of nonattachment, but I feel the objectively real phenomena must be recognized for what they are and then overcome.

Christianity tends to do the opposite: rather than ignoring psychic phenomena, it overreacts. Paranormal ability is either exalted and sanctified, as in the case of certain saints, or deplored and punished, as in the execution of thousands of "witches" and "heretics" during the Middle Ages. Given such precedent, it is little wonder that western society has lost or concealed any familiarity with nonordinary experiences of reality.

A healthier approach to the situation is possible. If we focus clearly on the fact that the purpose of awakening nonsensory states of consciousness is to further the evolution of humanity, we see that paranormal abilities can permit us to experience existence outside the realm of the five senses. Such experience, when kept in perspective, serves as an inspiration for further discipline and may be employed as an aid in freeing oneself from the bonds of the physical plane.

One major feature of paranormal experience is that the limitations of three-dimensional space-time may be transcended. One type of simple clairvoyance, for example, is to "see" an event taking place in the next room (consciousness passes through space, as it were, permeating the walls of the room and apprehending the object directly). Many people can accept this type of evidence quite easily. But examples of precognition exist which indicate that clairvoyance can also take place in regard to future or past events. The rational intellect has more difficulty dealing with this type of evidence and tends to ignore it—probably because the notion of forward-moving linear time is very deeply rooted in sensory consciousness.

Precognition nevertheless occurs. In the spring of 1972, I was scheduled to give a series of lectures in the United States and western Europe. After a lecture in Rome, my wife and I were supposed to fly back to Tokyo via Tel Aviv and Bombay. A week or two before my departure, I had a clear presentiment of impending danger at the end of the trip. I canceled all my lectures in Europe, deciding to go only to the United States, and I postponed my departure from April to May. On the final leg of the tour, which called for us to get from St. Louis to San Francisco, we made reservations on a flight for San Francisco via Denver. Then I again sensed that some mishap would befall the plane. So we canceled the reservation and took a more roundabout route via Phoenix and Los Angeles. When we arrived in San Francisco, we read in the paper that the plane we were supposed to have taken had been hijacked by Black Panthers in order to force the release of Angela Davis, who had been sentenced that day. Moreover, two weeks after my safe return to Tokyo, Japanese guerillas attacked Lod Airport and killed a number of people. If I had followed my first schedule, I would have been there on that exact day and might have been involved in the tragedy.

This sort of precognition has been with me all my life, but manifestations were sporadic and uncontrolled until I dedicated years to the practice of meditative discipline and other religious exercises: since then, they have become largely voluntary. The early paranormal phenomena that I experienced through nonsensory modalities of perception/consciousness were visual and cognitive. They were also passive, in the sense that I was able to gain various kinds of information about the external world, but I was not able to exert a direct influence on a given object or event. As centers of paranormal consciousness awakened, I gradually developed the ability to effect changes in the external environment, without using sensory means. These abilities, though they should never be used for selfish purposes, can be a valuable and powerful means of helping other people, particularly in the realm of healing.

One day, for instance, I received a telephone call from a friend living in Kyoto, about 500 miles away. His child was sick, with symptoms of high fever and dehydration, and had been confined to bed unconscious and unable to eat or drink anything for three days. I wanted to do anything I could to help the boy, so I went into a state of samadhi and tried to send energy in his direction. After a while, I became aware of his exact location, as if my energy transmission were operating like a radar tracking device. I then felt a strong, clear sensation that my energy connected to his, and somehow knew that he would be cured if I transmitted energy to him. After a while, I realized that a sufficient amount of energy had been given, and I came back to the ordinary state of consciousness, checking my watch as I did so. The next morning, the child's mother called to tell me that he had regained consciousness the night before. I asked her what time that had been, and the time she stated was the same time that I felt that I had disconnected from him. The boy's temperature soon returned to normal and he was able to accept liquids. After a few days the child recovered completely.

In such a case, my subjective awareness is that some form of subtle energy enters my body from the outside and flows through me, as though I were a valve or conduit. In order for this ability to function, I must empty my mind of any sense of personality or individuality. I accomplish this by following a psychological process similar to that described in chapter 2. As long as the individual consciousness can be suspended, the valve remains open and the healing energy is free to pass through to the suffering person. Physical distance does not seem to alter effectiveness.

I mention these examples in order to raise a few questions that have concerned me for a long time: How, in scientific terms, is all this possible? What are the underlying mechanisms by which such phenomena operate, and how can they best be studied so as to provide informative answers? My interest lies in attempting to verify and describe nonphysical dimensions of existence, particularly in terms of their relationship to the physical world. Paranormal phenomena, important in and of themselves, comprise a valuable field of exploration in that they appear to bridge the physical and nonphysical dimensions of existence. Through controlled experiments, they can be brought into the realm of objective investigation: moreover, they act as indicators that dimensions of nonphysical reality do exist, and, further, that the various dimensions are somehow connected or interpenetrating.

Materialists assert that the physical dimension is all there is: they hold to the axiom that a physical event must have a physical cause, and must occur according to set physical laws which govern nature. But if it were possible to establish, through strict scientific inquiry, that an observable phenomenon somehow defies generally accepted physical laws, the existence of a nonphysical dimension of nature would have to be hypothesized.

Parapsychology is one branch of science that has dealt with nonsensory modes of perception and activity. This field stemmed from psychical research, which arose in the late nineteenth century as a reaction against the dominant dictates of materialism. A steadily growing number of prominent scientists and thinkers became dissatisfied with the doctrine that matter is the only reality and that everything in the world, including the human mind, can be explained in terms of matter. This notion categorically denies the claims of the religious sphere that at least some aspects of human nature transcend the limitations of physical mass and energy. It also completely ignores the class of spontaneously occurring psychic phenomena.

These blatant discrepancies led some of the leading intellectual lights of the West, particularly in England, to organize themselves into societies for the purpose of exploring psychic experiences, verifying their occurrence, and thereby demonstrating that the human experience holds more than materialistic science has been willing to concede. In 1882 Frederick W. H. Meyers, a renowned classical scholar, and Henry Sidgewick, professor of philosophy at Cambridge University, together with other leading British scholars of the day, founded the Society for Psychical Research (SPR). Two years later a psychical research society was established in America, and in the years that followed, others were set up in many nations throughout the world. These organizations did not approach psychic phenomena from a broad perspective; they did not try to determine what such phenomena mean in terms of the nature of consciousness or reality. They attempted only to prove that psychic phenomena exist.

By and large they were unsuccessful. Though it accumulated a vast amount of information, the SPR was unable in the end to convince a significant portion of the scientific community of the existence of psychic phenomena. Case studies of psychics constituted the bulk of the material in their investigations, and scientific verification of subjective accounts is beset with difficult theoretical and procedural problems. The most significant drawback to this methodology, when used with most psychics, lies in the spontaneity of the phenomena under investigation. The research subjects whom SPR studied were not mystics who had undergone long years of discipline, but mostly mediums; though they did have some degree of paranormal ability, they themselves were unaware of the reason for it and were not in voluntary control. They were unable to repeat paranormal activity on demand. This presented a major problem, because replication is the cornerstone of the scientific method: unless a given experiment can be duplicated, its findings are judged unverifiable, hence suspect. According to the principles of science, such findings cannot be considered valid contributions to the store of human knowledge.

Another central difficulty with the work of the SPR was that mediumship, the phenomenon that the SPR chose to emphasize, posited the existence of spirits, rather than nonphysical laws of nature, as the explanation for psychic phenomena. But this hypothesis could not be satisfactorily verified. In fact, although the original aim of psychical research was to verify scientifically the occurrence of phenomena that were inexplicable in terms of physicalist laws of nature, participation in the research came to imply the acceptance of certain metaphysical assumptions biased in favor of spiritism. Largely because of these problems, the vast amount of evidence accumulated by the SPR, though it persuaded the participants of the occurrence of psychic phenomena, did not sway a skeptical scientific world.

It became clear early in this century that an experimental method was needed by means of which psi—as psychic perception and power had come to be called—could be tested repeatedly under controlled conditions and evaluated quantitatively. Various attempts were made, but it was not until the 1930s that a methodology was devised that transformed psychical research into an experimental laboratory science. Dr. J. B. Rhine,[3] who had originally set out to prove the immortality of the soul, sifted through and eliminated prejudicial metaphysical assumptions from the experimental method and the analysis of results, and rigorously defined the phenomena under review. Clairvoyance, telepathy, precognition, and GESP (general ESP, a combination of the first three) were fixed as the objects to be studied. Similarly, carefully controlled laboratory techniques were designed for testing PK (psychokinesis).

To study ESP, Rhine and others devised the now famous ESP cards, a pack of twenty-five cards, each printed with one of five simple symbols; there are five cards of each type in every pack. The use of these cards in standard runs of twenty-five trials represents a repeatable, controllable laboratory procedure. Anyone can serve as the subject of experimentation; moreover, the experiment is designed so that the results can be evaluated mathematically.

The ESP-testing technique Rhine devised was elegantly simple. By shuffling the cards thoroughly, out of the subject's sight, the experimenter converts the pack into a sequence of twenty-five random events, in unknown order. As the experimenter pulls out each card, the subject tries to identify it using ESP; in each trial, by choosing one among five ESP symbols, the subject has a one-fifth probability of being right. According to chance, therefore, the most likely score for the entire run of twenty-five cards would be five cards. Statistical theory holds that when the odds against the chance occurrence of a given score are greater than 100 to 1, the phenomenon can no longer be dismissed as being due to pure chance alone; another explanation must be sought. Therefore, if the number of hits scored in an ESP test are significantly greater than chance expectation, we can assume that the subject was able to gain information about the card symbols nonsensorily. The safeguarding conditions imposed on the test procedure (those preventing any sensory perception or communication of the card symbols), coupled with results that make it unreasonable to assume that chance alone was involved, leave no alternative explanation.

After establishing his ESP procedures, Dr. Rhine proceeded to devise a new approach to the study of PK, the power of mind over matter. He initiated exploratory investigations involving dice-throwing in an effort to test the influence of human will upon the behavior of matter. As the experimental procedures were further refined, dice-tossing machines eventually replaced human hands. By computing the results on the basis of statistical laws of probability, it was once again possible to determine mathematically that chance alone could not account for the results obtained. (In one series of 651,216 die throws, the odds against the total number of correct answers was 10^{115} to 1!)

Several decades of parapsychological research conducted by Dr. Rhine, J. G. Pratt, and many others at institutes and universities throughout the world, have demonstrated in the laboratory that a form of perception can occur even in the absence of both a recognizable stimulating energy emanating from the target object and a sensory organ specifically differentiated for the perception of that object. Similarly, the irrelevancy of space (in distance ESP tests) and time (in precognition experiments), and the fact that external physical systems can be subjected to direct mental influence by PK, argue for the existence of a nonphysical or "mental" energy in humanity and in nature. Moreover, all these findings were accomplished by an analysis free of unverifiable metaphysical hypotheses.

However, the findings obtained through the quantitative method provide no information as to the mechanisms involved: they suggest that psi phenomena are not produced by physical means but do little to explain how they *are* caused. Possibly the only major positive characteristics of psi manifestation discovered thus far are that ESP functioning seems to be related to and is influenced by the unconscious, and that emotion plays an important role in the subject's ability: levels of interest and motivation directly influence the scores, both positively and negatively. Even though the subject's emotional state has been established as a condition affecting the appearance of ESP and PK, the nature of the actual connection is not known. Again, we run into the frustrating limit of science. Nonsensory phenomena have in no way been directly apprehended by these experiments. Only the end products, phenomena which appear to have been produced by the actions of a nonphysical mechanism, have been observed.

In the search for a more informative body of evidence, certain scientists, particularly in Europe, have continued work along qualitative, rather than quantitative, lines. They use the case study approach, but treat the data with modern psychological criteria which were not available to the early psychical researchers of a century ago, and they incorporate the principles of science as much as possible.

The work of Dr. W. H. C. Tenhaeff, former director of the Parapsychological Institute of the State University of Utrecht, Netherlands, is representative of the European school of qualitative research.[4] Dr. Tenhaeff views the realm of psi as a branch of psychology, and, as an advocate of the introspective method of psychological analysis (a technique of self-observation originally used to monitor the higher psychical processes of normal thought and volition), he has attempted to describe parapsychic consciousness. Tenhaeff's subjects have consisted primarily of paragnosts, people who repeatedly and at regular intervals acquire knowledge through ESP. After checking that the information received through nonsensory means is correct, Tenhaeff investigates the subjects' subjective interpretation of the manner in which thay came to acquire this knowledge. By carefully recording the data from repeated sessions with many paragnosts, and by comparing his findings with those of other scientists, Dr. Tenhaeff has gradually built up a foundation for a descriptive psychology of paragnostic consciousness.

For example, most of Tenhaeff's subjects reported—and were observed to enter—a state of reduced awareness, a state of introversion during which outside stimuli were momentarily blotted out. These states varied in intensity from subject to subject, ranging from light trance to deep trance and somnambulism. This observation fits in well with the yoga teachings described in chapter 2, which state that ordinary waking consciousness must be restrained from reacting to the external environment and quieted before nonphysical forms of perception/consciousness can manifest.

Tenhaeff has also shown how an individual's psychological make-up can influence the type of paranormal phenomena experienced. In many cases of precognitive experience regarding the death of royalty, for instance, the individuals appear, upon analysis, to be suffering from the so-called Oedipus complex. "Thanks to their clairvoyant powers," he says, they were able to "settle with the father image in actual, guilt-free identification with regicides." He relates the example of a female patient with a strong father fixation who had a precognitive dream in which she vividly saw the automobile accident that took the life of Prince Bernhard of the Netherlands, an event that actually occurred two days later. Tenhaeff adds that this case of emotion-laden precognition is by no means unique. All such cases show how the dividing line between the individual unconscious (specifically, a part of the individual unconscious which harbors much "emotional energy") has disappeared, if only momentarily, to unify with a larger reality.

Though the methodology may differ, both the quantitative and the qualitative approach attach primary importance to verifying the existence of psi and to determining the psychological mechanism of its manifestation. There is yet another approach—not a purely parapsychological one, however—which has become very popular in the past fifteen years: the holistic study of humanity from the standpoint of energy rather than consciousness.

One of the pioneers in this area was Dr. L. L. Vasiliev, Chairman of Physiology at the University of Leningrad. The Russians began parapsychological research in the attempt to prove once and for all that nonphysical energy, including the soul, could not possibly exist. Paradoxically, they ended up proving exactly the opposite.

The Russians began with the hypothesis that paranormal phenomena such as telepathy and psychokinesis were most likely caused by electromagnetic waves of either very high or low frequency, undetectable to the ordinary sensory apparatus. They likened the human body to a living radio and hypothesized that it might be able to pick up subtle electromagnetic waves by means of a yet unknown sensory organ or perhaps the brain itself.

To test this hypothesis, Dr. Vasiliev and a team of researchers conducted a series of experiments in which they attempted to demonstrate that telepathy could not occur if the sender and receiver were electromagnetically isolated from each other. They tried everything—lead-shielded rooms, Faraday cages, physical distance (1,000 miles)—but, to their astonishment, telepathy clearly could occur under such conditions. Ironically, they were forced to conclude that whatever is responsible for the direct transmission of information from mind to mind is not a part of the known electromagnetic spectrum, that telepathy does not depend on physical energy. This conclusion was reached in 1937, but Vasiliev was not permitted to publicize his findings until 1962, when his work, *Experiments in Mental Suggestion*, was finally published.

The study of nonphysical energy forms has been undertaken from many different directions by various types of researchers—physicists, psychophysiologists, physicians, electrical engineers. The main thrust of this research is physiological: it deals with energy fields in and around the human body and the relationship of these fields to the subject's mental state. Included in this field is the work of such scientists as Semyon and Valentina Kirlian, L. L. Vasiliev, Itzhak Bentov, E. Douglas Dean, William Tiller, Wilder Penfield, Shafica Kargulla, and a growing host of others. The independent findings of these scientists have shown that we are composed of more energy systems than was previously assumed.

My own research, discussed in the following chapter, has been concerned with the nature of the various types of relatively unexplored energy which form the human body, and the question of how these relate to consciousness, both sensory and non-sensory.

CHAPTER 4

The Mind-Body Connection

WHAT IS the nature of the connection between mind and matter, the non-physical and the physical? Do they exist separately? Does one cause the other, or vice versa? Which, if either, has the dominant influence?

Science has not yet successfully answered these questions. We do not even know where consciousness is located. Whereas some cultures have decided that consciousness emanates from the heart, say, or the belly, most modern scientists assume that consciousness is housed in the brain. Studies such as those of Dr. Wilder Penfield, however, have begun to show that the linkage between consciousness and the brain is somewhat tenuous. Penfield's research suggests that consciousness, rather than being located in the brain, may be an independent entity which functions "through" the brain.

Certain branches of science have gone so far as to say that consciousness, as a nonphysical entity, does not really exist at all. This is the basic tenet of the behaviorist school (favored among many leading scientists and psychologists from 1920 to 1960), which assumes that all mental functioning is essentially a neural process. One behaviorist theory holds, for instance, that the mind-body relation can be explained physiologically in terms of the reaction taking place among the centers of emotion and the autonomic nervous system in the diencephalon and the limbic system; another claims that all mental states can be regulated by stimulating or relaxing certain areas of the brain with psychotropic drugs. Of course, the body *does* influence the mind. But since it is possible to demonstrate that these physical determinants of psychological phenomena are not absolute, a position that identifies the two is untenable.

In general, the mystics view the mind-body relation as one in which both mind and body are aspects of one essential continuum, the unifying principle of reality. Consciousness, energy, and matter are interconnected phenomena. Modern research is finally beginning to produce evidence in support of this experiential insight. Research has shown that the mind can influence the body, can permanently alter the physiological balance presently considered normal by medical science, and can even influence the workings of another person's body.

The effect that the mind can have on the body seems a natural place at which to begin an investigation of consciousness, because the body can be monitored closely by available technology. At this stage, we can only gain information about the nonphysical by examining its influences on and manifestation in physical reality. Because nonphysical existence is most readily apprehensible to human beings as individual consciousness, and matter as the physical body, it seems logical to approach the problem by studying the effects that different mental states have upon the physical body. It is to be hoped that this type of investigation will also turn up clues about the deeper mystery of the exact mechanism that relates psyche and soma.

At the same time, this sort of inquiry offers an excellent opportunity to explore another tenet of the mystical world view: that whenever profound alterations of consciousness occur, resonating changes are produced in the body as well. There is a growing amount of evidence that indicates that as individual consciousness is transcended and unification with other, greater forms of consciousness occur, recognizable changes concurrently begin within the nervous system. Science must seek a detailed understanding of those changes.

To begin my scientific quest, I had to devise a research methodology that would combine a number of factors: the basis of the research was to be physiological, but its goal was psychological and parapsychological understanding. I wanted to explore not only the connection of individual consciousness to the body, but also the relationship between nonsensory states of consciousness and the physical world.

The exact scientific investigation of parapsychological phenomena is beset by problems. Experiments must be repeatable and the results analyzable. The American parapsychologists (including Dr. Rhine) devised experimental schemes that met these terms, proving that some dimension of mind that is able to transcend the limitations of sensory transmission, time, and space does exist. Under strict laboratory conditions, it was shown that consciousness is able to operate independent of physical channels.

However, the need to restrict the operation of psi (to guessing the identity of ESP cards, for example) so that it fits into the quantitative system greatly lessens the degree and quality of the phenomena under review. There is evidence that ESP, for instance, is somehow connected to the unconscious and, like most unconscious processes, cannot—in most individuals—be turned on and off at will. But some people seem to exercise a fair amount of conscious control over their ESP functioning. Such people are rare enough so that few scientists have been able to find large enough samples to test quantitatively. However, these are the individuals who, when studied qualitatively, can help produce a more comprehensive explanation of the aspects of consciousness that transcend the normally assumed limitations. Though a few scientists began sporadically in the 1930s to test Indian yogis and other people with an extraordinary amount of control over their bodies or consistent psychic abilities, this type of research has not been done systematically until recently.

In the late 1950s the idea first occurred to me that the quantitative, qualitative, and physiological approaches could be combined in one procedure. I decided to begin by gathering a group of subjects who had undergone some form of ascetic or spiritual discipline (such as that described in chapter 2) and felt that they had experienced some degree of unification as a result. Then I compared these individuals (group A), by means of standard parapsychological and physiological testing, to two other groups: the first (B) comprising people who had been meditating for some time and were subjectively experiencing changes in physical and psychological functioning which they attributed to the discipline, but who had not yet undergone any distinct breakthrough into other dimensions of consciousness; the second (C) a random sampling of students from the university, friends, anyone who was not particularly interested in the spiritual or psychic realms of life (the only requirement was that these subjects be in good health). By comparing the three groups, I hoped to delineate any statistically significant differences among them and thereby to verify whether the mystical experience recorded throughout the centuries had any scientifically valid basis in reality.

A group of meditators, who had been meeting together regularly since 1953 under the auspices of our shrine, proved to be ideal subjects. This group met twice a month for meditation and discussion, and the individuals practiced daily in their homes. Some had been with us since the inception of the meeting; others were relative newcomers. The male-female ratio was fairly balanced. They included Shintoists, Buddhists of all sects, and Christians.

Each member of the group, which numbered over one hundred, was in a different phase of the meditative process. Six who had been meditating for a long time (ten or more years) claimed to have experienced unification with non—sensory-dependent states of consciousness. There were other individuals who had been meditating for as long, but these six in particular were thought to have transcended the normal condition, and each had shown definite ESP or PK capabilities. One was my spiritual guide, Odaisama, who had recorded a substantial number of precognitive experiences that had manifested in reality. She had also performed many healings which seemed to defy medical explanation. Another was an aging Zen master who had spent most of his life observing strict Buddhist practices. Another was the founder of a widespread religion here in Japan, who also had had a great many precognitive and healing experiences.

During the first eight years of the meeting, about twenty members of the group had experiences apparently attributable to contact with nonordinary dimensions of reality. There are quite a number of symptoms, both physical and psychological, which are traditionally said to accompany this process. For instance, during meditation there may occur a sudden, all-enveloping feeling of deep peace which temporarily immobilizes one. The breath suddenly slows down and the body may become very hot. Occasionally the body will start to shake uncontrollably. In general, dietary and digestive habits may change drastically. Emotions often become unstable. Another common manifestation is the onset of "psychic experiences" or a distinct increase in their occurrence and clarity.

One very intelligent young woman, the daughter of a chemistry professor, had joined the group seeking some peace and stability. She was depressed, having recently divorced her husband. Her highly rational upbringing made her very skeptical about psychic powers, and she clearly voiced her criticism whenever the topic came up. After five years of meditation she began—to her chagrin and amazement—to have unmistakably precognitive dreams, often concerning her relatives. In one her nephew became seriously ill, which he did in actuality one month later. In another she saw her mother leave home suddenly on an unexpected trip because of some problem to do with her brother. Five days later, this also happened exactly as she envisioned it.

Among these twenty or so people, others had just begun meditating but had experienced sporadic psychic events since childhood. I hypothesized that such psi-latent personalities, those with seemingly inborn psychic tendencies, might possess an innate ability to dissociate from the noise of active awareness and receive information through nonsensory channels. Through personal interaction and interviews, I became aware that the psi-latent types showed many intrinsic physical and personality traits similar to those of people in the midst of the meditative process. They dreamed a lot, were usually high-strung and emotional, often had some form of stomach or intestinal difficulty, and had a strong yearning for mystical experience (curiously, William James, in *Varieties of Religious Experience*, describes those who are likely to have religious experience in some of the exact same terms). The similarities were so consistent that even if people had not been meditating for a long time but exhibited the above characteristics, especially that of spontaneous psychic experience, I included them in group B.

The groups were classified not only according to the personal experiences of the subjects, but also through the way I perceived them and the way we perceived certain events collectively. For instance, at times, during the group meditations, I would become aware of specific nonphysical (disembodied) presences in the shrine. If these manifestations had any objective reality (in the sense that people with normal sensory vision agree on the word "red" to represent a certain color), it could be expected that any person possessing the same type of nonsensory vision as mine would be able to see the same thing. To determine that my awareness was not just subjective illusion, I might ask the group, when the meditation was over, what, if anything, they had perceived extrasensorily. In quite a number of instances, the same presence was described by various group A individuals; each would add another facet to the portrait of that which was perceived.

An empirical mystical approach was also used to classify subjects. According to yogic teaching, there are seven chakras (or centers of energy and consciousness) located along the spine, and each, when awakened, stimulates psi abilities and has a characteristic color which appears as a surrounding ring of light when observed through nonsensory perception. In many Hindu sects, competent gurus are able to perceive immediately which of their disciples' chakras are active by using their own power to view the colors emanating from the disciples. A leader of the Ramakrishna sect, for example, told me that this method is still used to determine the extent of a disciple's development; through it the disciple's position in the sect is assigned.

Using this method within the group, relying on my own nonsensory perception and that of the members of the group who had similar vision, I gathered information which I later used in forming the research groups. Subjects alleged to have active heart chakras—which, according to Hindu doctrines, should be indicative of relatively advanced ESP and PK abilities—were placed in group A, subjects with activation in any of the three lower chakras (correlated with lesser paranormal abilities) were placed in group B, and those with no perceptible chakra activity were placed in group C.

The A-B-C classification has one unique feature that is of central importance to an understanding of my research. Up to this point in the history of science, all hypotheses and theories have begun with the sensory observation of phenomena. Scientists observe some event, such as an apple falling from a tree, and then go on to formulate a hypothesis from the initial observation. By quantitatively testing the hypothesis, the scientist finds the general principle underlying the event (e.g., "Every object in the universe attracts every other object,") and calls it a law of nature. My research methodology follows the same line of reasoning, but the starting point is different. Rather than beginning with sensory observation (one type of empirical knowing), I start, in part, with nonsensory observation (another form of empirical perception).

Physical sciences attempt to describe the physical universe

by using physical means. There is no room in this scheme for an investigation of the alleged nonphysical dimensions of being. To initiate an investigation into the nonphysical areas of life, then, I felt it necessary to begin with nonsensory observation (my own as well as those of the many research subjects who appear to have the same abilities), and the observations repeated in traditional mystical literatures.

Over the years, the number of subjects I tested grew beyond the original hundred into the thousands. Generally, I would insert new subjects in the A-B-C classification. Since group A subjects are relatively rare, in order to gather pertinent data I occasionally had to make special efforts, such as a trip I made to India in 1963 to examine some twenty well-reputed yogis. Fortunately, a sufficiently large number of psychically developed individuals have agreed to cooperate over the years, so that the A-B-C comparison could be maintained throughout.

One of the early experiments utilizing the three groups was a series of standard ESP card tests on 91 subjects carried out in 1961 to gauge the validity of my classification system. Group A consisted of 12 subjects, B of 25, and C of 54. When the total results of the scores of each group were analyzed, a remarkable difference was found among the three. Group C (259 runs), with an average score of 5.116 correct guesses, showed a probability of 0.3524, which lies within the realm of chance. Group B (216 runs) had an average score of 5.412, the probability being 2.48×10^{-3}, which means that ESP was statistically demonstrated by this group. Group A (243 runs) averaged 6.1119, giving a probability of 1 in 10^{-15}. In other words, there is only one chance in a trillion that group A could have obtained such scores by chance.

Thus, a significant difference appeared among the three groups, and because of the way the experiment was designed, the only plausible explanation for the difference is the degree to which ESP is manifest in the members of each group. Group A— who could be assumed to possess the greatest degree of paranormal development, if the method of selection was correct— indeed showed the greatest ESP ability; group B came in second, and C third. Not only do these results suggest that the method by which the groups were chosen was valid, they indicate a definite link between spiritual discipline and ESP function, in that the group of subjects who had practiced meditation for the longest time showed the greatest ESP abilities. This experiment was repeated in various forms over the years, with similar results.

Dr. Rhine and other quantitative researchers agree that ESP functioning is connected to the unconscious. This connection was deduced, in part, from the fact that subjects themselves are by and large unaware of how accurately they are hitting the target object. This has led other researchers to hypothesize that ESP function might be directly reflected in the autonomic nervous system, which is the physiological counterpart of the unconscious.

The human nervous system coordinates the activity of several million nerve cells into one cohesive functioning unit. It is divided into three major parts. The *central nervous system*, composed of the brain and spinal cord, communicates through the *peripheral nervous system* to all parts of the body. The *autonomic system*, which is largely independent of the central nervous system, regulates many vital functions of the body.

The autonomic nervous system has two components. The *sympathetic nervous system* serves a stimulative, excitory function. It prepares the body for emergency, causing blood vessels to constrict, the heart to beat faster, and the pupils to widen. This activity is countered by the other half of the system, the *parasympathetic nervous system*, which inhibits activity. Parasympathetic excitation induces the relaxation of blood vessels, slow but strong pulse, and a general feeling of well-being. The sympathetic system comprises two cords of nervous tissue, which run down either side of the spinal column. Each cord has separate masses of nerve cells, known as ganglia, interspersed along its length. The largest of the ganglia is the solar plexus, located below the diaphragm. Another is found in the lower part of the abdomen, a third is near the heart, a fourth is located at the throat. The main vehicle of the parasympathetic system is the vagus nerve that branches off from cranial nerves to its own system of ganglia which are located near target organs.

The two halves of the autonomic nervous system function together to regulate and harmonize the manifold life-sustaining functions of the body. We may eat an apple without difficulty, but imagine if we had to attend consciously to every process necessary to digest it! Fortunately, we can forget about the apple as soon as we have eaten it, and trust that unconscious processes and the autonomic nervous system will do the rest.

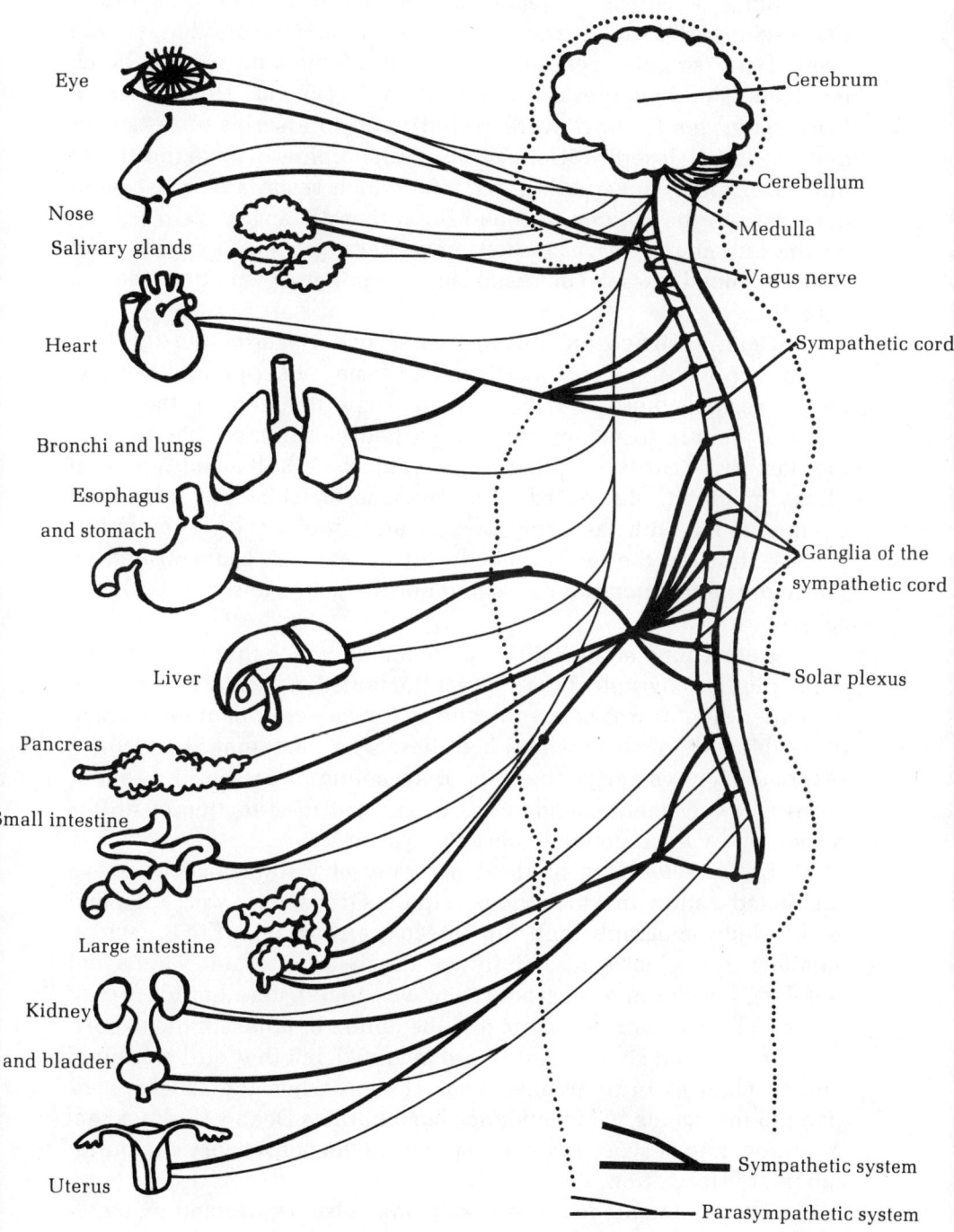

Figure 1. The autonomic nervous system and the body's organs.

Some scientists conjectured that, if ESP is connected to the unconscious, ESP might register in the autonomic nervous system even if the subject were not conscious of the information being received. This hypothesis was tested by E. Douglas Dean at Duke University;[1] his findings were reconfirmed in a series of tests conducted at St. Joseph's College in Philadelphia. The experiment utilized a plethysmograph, a machine which records increases and decreases in the volume of blood flow; these changes are triggered by the autonomic nerves, which have been proven to expand and contract blood vessels in response to emotional and unconscious stimuli.

Dean designed an ESP test using twenty cards. On the first five he wrote names picked at random from a telephone directory; on the second, those of friends of the experimenter; on the third, those of people to whom the subject had close emotional ties; on the last five cards, nothing. The subject was hooked up to a plethysmograph, the recording mechanism of which was placed in another room, with the experimenter and an observer. The experimenter shuffled the deck, picked up the top card, concentrated on the name, and attempted to communicate extrasensorily with the subject.

Dean found statistically significant evidence that the baseline of the plethysmograph of the subject fluctuated more markedly while the experimenter was concentrating on the names of people to whom the subject had close emotional ties. The fact that the subject responded physically suggests that communication did occur between experimenter and subject, and that the function of ESP is registered by the autonomic nervous system.

Dr. Charles Tart of the University of California–Davis has conducted a study in which the percipients (receivers) were attached to a plethysmograph and an electrocardiogram (ECG), which monitors the electrical activity of the heart muscle. The agent (sender), located in a separate room, was administered mild electric shocks. The receivers were not told the nature or object of the experiment and did not know agents were involved, but they still exhibited sudden changes in their physiological signs when the shocks were given to the agents.[2] This evidence corroborates Dean's findings that the autonomic nervous system responds to the nonsensory communication of information.

The autonomic nervous system may also be affected by meditation. To verify whether it is or not, and in what way, I sought first to demonstrate the way in which a normal group of subjects might be able to influence autonomic function through the power of mind; I then moved on to investigate how individuals with extensive meditation experience compare in their ability to do the same.

In the series of experiments I designed, I used, in addition to the ECG and plethysmograph, a GSR (galvanic skin resistance) recorder, a machine that monitors sympathetic nervous function by detecting changes in the electric potential of the skin (this is one of the body's responses, by the way, which is monitored by "lie detector" machines). I also used a pneumatograph (a device which monitors the frequency and pattern of respiration), and an electroencephalograph, or EEG.

The first fact I needed to verify was whether consciousness is able to affect the functioning of the autonomic nervous system at all. Years ago, when this simple research was done, it was not yet clearly recognized that the mind, through a simple act of will, could influence the autonomic nerves. Later the sophisticated technology of bioelectric measuring tools would enable scientists to monitor subtle physiological changes previously undetectable.

In one experiment I compared the physiological differences concomitant to mental, arithmetic calculation to those present during one-pointed concentration; I concluded that mental calculation causes excitation of the sympathetic nerves (evidenced by an increase in shallow respiration), whereas one-pointed concentration relaxes the system, leading to a slower, deeper, longer respiration.

In another experiment I examined the difference concomitant to sustained concentration in members of groups A and C. I asked subjects to focus their awareness for a certain period of time at a point between their eyebrows. For control purposes I took plethysmograph, pneumatograph, and GSR readings for four minutes while the subject was lying quietly. After a one-minute wait, I started the measurements again, running them for one minute before the subjects were asked to begin concentration. The readings were continued while the subjects attempted to maintain concentration for three minutes. The results of this experiment showed, again, that mental concentration does affect the subtle physiological processes of the body. Most group C subjects showed an increase in respiration rate. Group C's rate, averaging 16.8 breaths per minute in control (the norm is 16), rose to an average of about 23 per minute during concentration. Group A, on the other hand, evidenced a sharp decrease in respiration rate during the experiment, averaging about 15 during control and 5 per minute during concentration.

Statistically, group A showed a greater degree of change during concentration than group C. This difference indicates that group A seems able to exert more influence over their bodies than group C. A direct relationship is likely here between the degree to which group A subjects have developed their access to nonsensory levels of consciousness and the amount of control they are able to exert over their bodies, a hypothesis that was borne out time and again in later experiments.

Such results are complemented nicely by a number of psychophysiological investigations into meditation which have been carried out by other researchers (e.g., Wallace, 1970; Wenger and Bagchi, 1971). All of these studies show that meditators undergo a change in pulse and breathing rates during the practice. Interestingly, however, some researchers conclude that respiration is slowed during meditation (that is, that the parasympathetic is in control), whereas others find an increase in metabolic rate and strain of the sympathetic system. My own studies suggest that the changes in physiological function depend on the length of time the subject has been meditating and the particular method used.

The effect that mind and body exert on each other has been demonstrated repeatedly in biofeedback studies. Biofeedback technique is based on the supposition that the will, aided by subtle biological feedback, is able to influence the autonomic nervous system and reprogram reactions to stress. Elmer and Alyce Green, in their book, *Beyond Biofeedback*,[3] formulate the principle that "Every change in the physiological state is accompanied by an appropriate change in the mental-emotional state, conscious or unconscious, and, conversely, every change in the mental-emotional state, conscious or unconscious, is accompanied by an appropriate change in the physiological state." Their research, as well as the data collected by many other biofeedback researchers, shows that the conscious mind can gain a formerly unsuspected degree of volitional control over this system of cause and effect—that mental will can initiate changes in the subtle body processes leading to self-regulation and the release of stress in the system. These subtle body processes are those controlled by the autonomic nervous system and are therefore unconscious, but the conscious mind can link up directly to the unconscious and cause it to function in a desired manner. They hypothesize that the process occurs in the following way: an individual can "(a) desire and visualize a certain kind of physiological behavior (this is a conscious cortical process); (b) plant the idea in the unconscious, the earth of our psychological being (physiologically, the sub-cortical area); (c) let nature take its course; and (d) reap mind-body stability." By monitoring the electrical changes which the body undergoes when the will thus initiates a change in autonomic nervous function, the Greens, and other researchers, have clearly demonstrated the the mind can influence the "unconscious" processes of the body.

The power of consciousness to affect supposedly automatic bodily processes is in itself quite amazing, but it is even more interesting that consciousness can demonstrably induce changes in physiological functioning which lie totally outside the normal range of autonomic activity.

In a healthy subject there is a balance in the function of the two halves of the autonomic system, and a certain range of activity to the processes they control. For instance, the heart beats within a spectrum of approximately 70 to 180 times a minute (70 at rest, 180 during intense physical exercise). Anything above or below this range is considered abnormal and, in most cases, pathological. If the autonomic system is forced to work beyond its range, if some stimulus is given that it cannot react to within its normal span of activity, pain or some other danger signal will result. But certain people—notably Indian yogis—have been observed to push their hearts up to 300 beats per minute with no apparent damage. Parapsychological research and mystic teaching both assert that consciousness can surpass its physical (sensory-dependent) limitations. Apparently, so can the body.

Figure 2 represents the amazing ECG of an Indian yogi taken at the Lonavala Yoga Research Institute by a professor of medicine at Bombay University. The yogi was instructed to attempt to stop the movement of his heart. About 30 or 40 seconds after measurement was begun, the intervals between pulses became 1 or 2 seconds. After 1 minute, the ECG shows a 5-second interval where the heart completely stopped. Clearly, the instinct of the heart to beat, one of the most basic and automatic instincts we possess, is not totally under the control of the unconscious. Consciousness can gain control of a phenomenon even as deeply rooted as this one is in the subtle func-

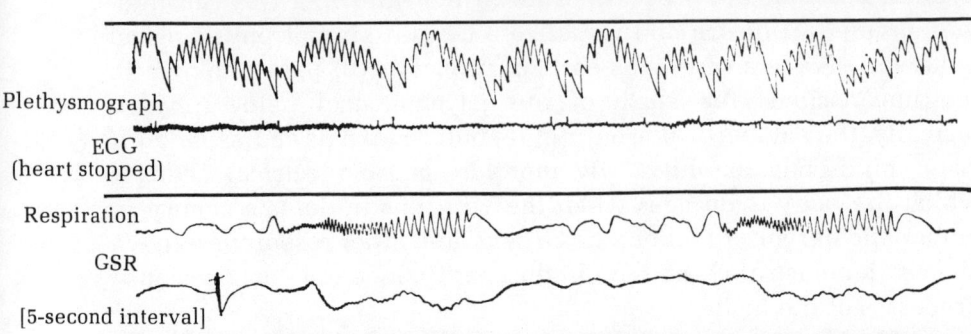

Figure 2. Readings taken on an Indian yogi stopping the movement of his heart.

Figure 3. Readings taken on an Indian yogi exercising the "rapid breathing" (bastrica) technique.

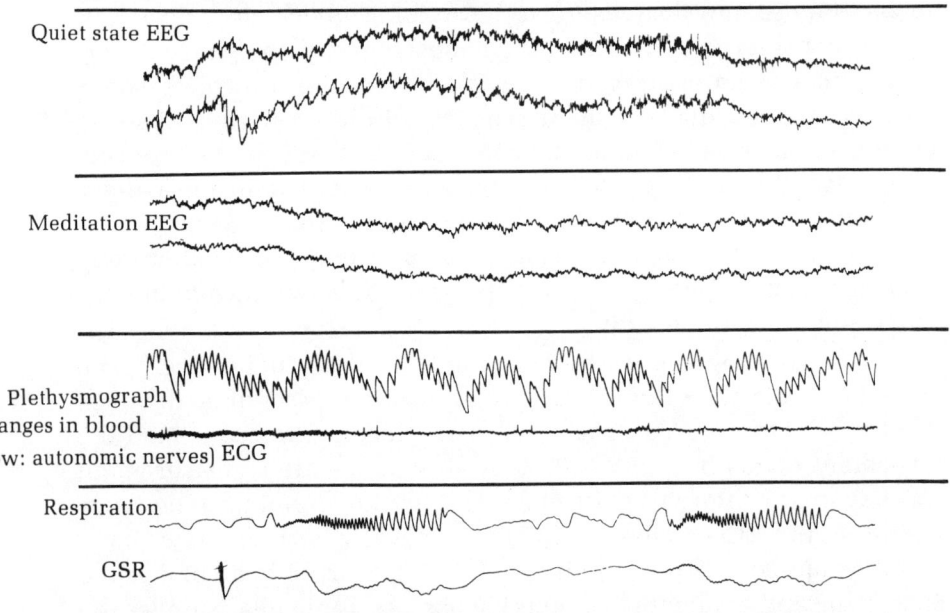

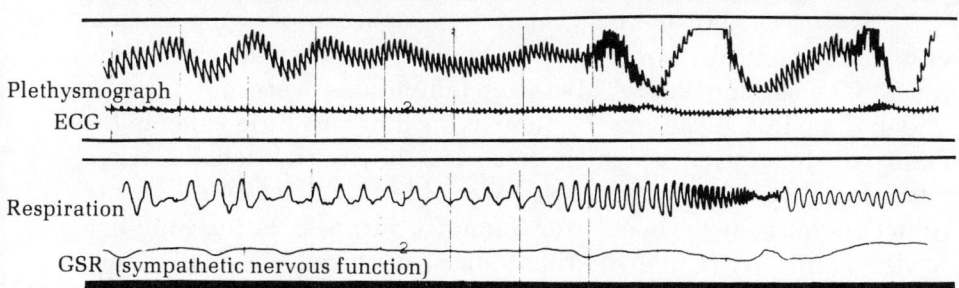

Figure 4. *Bastrica performed by a novice.*

Figure 5. *Agent-Percipient test with Tony Agpoa acting as "sender": before, during and after concentration.*

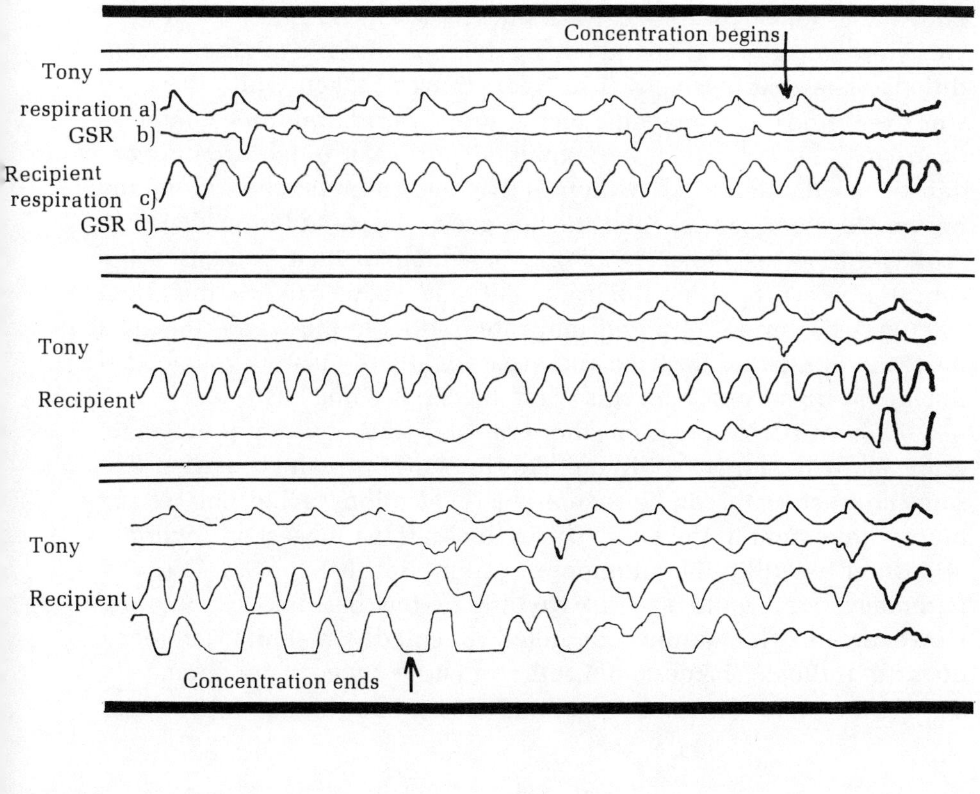

tioning of the physical realm. However, the scientific community has so far largely ignored information such as this, considering it—without examining it—to be out of the range of possibility.

Unfortunately, the next set of data (figure 3), showing unusual autonomic control, was recorded under crude conditions. Until recently, most spiritually advanced individuals were unwilling to undergo the investigations of scientists, as they saw little value in the materialistic approach which scientists have traditionally taken toward reality. Most of the yogis I was able to contact in India were reluctant to travel, especially for scientific purposes, so instead I had to visit some very remote districts. Once I ventured to a small village so far removed that it was still plagued by wildlife; the night before I arrived, a tiger had devoured a number of people. Under such circumstances, I was only able to take two portable two-channel EEGs with me.

Sri Satyanarayana, who claimed to have attained other levels of consciousness by awakening of kundalini energy, lives in a small village that is about a ten-hour drive from the University of Rahjasthan. Recently he has shown increasing interest in sharing his knowledge and experience with sincere scientific investigators: he was kind enough to let me take his polygraph under a number of conditions. The readings shown in figure 3 were taken while the guru was executing the pranayama technique of rapid breathing known as *bastrica* ("the bellows"). An extended period of quick breathing is dangerous if attempted without sufficient training: the autonomic nervous system cannot tolerate the excessive stimulation. Bastrica greatly increases the oxygen intake, which in turn decreases the internal pressure within the brain. Satyanarayana can breathe about 200 times per minute (the normal rate is 16). On the graph, the guru moves in and out of bastrica and breathes slightly faster than normal during the 18-second intervals when he is not doing bastrica.

The fifth tracing on the graph is that of the plethysmograph, showing large yet rhythmical fluctuations in the baseline. The same characteristic can be seen in the respiration tracing, and we see much excitement in the GSR. These fluctuations suggest a functional abnormality within the autonomic system. But if this were a case of pathology, we would see no rhythm in the baseline fluctuation. However, this fluctuation (one noted repeatedly in similar subjects) does not indicate sickness, but rather a wider range of function.

The state described by this data, one induced by rapid

breathing and concentration, differs from physiological measurements taken of subjects in contemplative meditation. In Zen meditation, for example, which consists mainly of strict sitting meditation and does not employ such breathing practices, pulse and respiration tend to slow down. Thus, the physiological patterns produced by the higher states vary somewhat, according to the method used to reach them. The important point, for our purposes, is that the autonomic nervous function is altered as a result.

One of Satyanarayana's disciples, an assistant professor of electronics at the Institute of Technology in Kakinada, was also measured while he performed bastrica, which he had been practicing for less than a year. Interestingly, there is no rhythmical fluctuation in the baseline of the plethysmograph; his GSR shows no excitement of the sympathetic nerves. These are valuable data, as they unmistakably indicate differences between the guru and the novice.

The above information shows us two things. First, we see that by making normally unconscious physiological processes conscious, one is able to alter subtle body processes at will. This alteration occurs within the normal range of physical activity. Second, we see that the range of autonomic activity itself can be widened, perhaps as a result of surpassing the limits of individual consciousness.

If mind has the potential to influence matter to such a degree, what is the possibility that one person's mind could consciously affect the body of another individual? I designed another experiment to explore this question. The agent (sender) and percipient (receiver) were located in different rooms, electrostatically isolated by lead shielding (no electrical energy fields could pass between them). Neither the agent nor the percipient were told anything about the test procedure beforehand. Both were hooked up to the physiological testing equipment. Then we asked the agent to concentrate on "sending energy" to the percipient in the next room. Measurements were taken throughout, to determine whether any significant changes occurred under test conditions. In many of the tests, significant changes were found. Some of the results were absolutely startling, as in the case in which Tony Agpoa, a well-known psychic surgeon from the Philippines, acted as the agent.

I first met Tony on a trip I made to the Philippines with Harold Sherman, a renowned psychic, in 1966. We went there to investigate the incredible rumors that certain people were able, without the aid of surgical instruments, anesthesia, or antiseptic, to make incisions in the human body, remove diseased tissue, and close up the wound without leaving a scar—a procedure that has been given the name psychic surgery. We observed several such operations, many of which appeared to be legitimate: Sherman describes them in his book *Wonder Healers of the Philippines*.[4] Later, I would return to the Philippines and do biochemical tests on the blood and tissue removed from actual patients (I'll discuss these findings later), but on the first trip I just brought along a portable polygraph.

Tony agreed to let me examine him. I wanted to check whether he showed any functional abnormality in the autonomic nervous system. At his home one day, I hooked him up to the equipment, and we ran a control. Everything was working fine. Then I asked him to concentrate in the same way he does when performing psychic surgery. Suddenly there was a loud hiss from the machine, and the equipment blew out.

Because my examination was so unsuccessful, Tony agreed to visit the Institute in Tokyo. He consented to act as an agent in the above experiment. We instructed him to try to send his "healing power" to a subject lying down in the next room. He was given an electrical signal to begin and another to stop. Figure 5 presents data from one of these tests.

The top two lines (*a* and *b*) represent Tony's respiration and GSR, respectively. The bottom two lines are those of the percipient. The moment Tony began concentrating on the subject, her breathing became faster and irregular; the moment his concentration stopped, it started to slow down. As for the GSR, it took somewhat longer for the subject's sympathetic nerves to become strained—about 20 or 30 seconds, but the fluctuations became quite forceful between 40 and 60 seconds.

Tony's data show opposite reactions all along. After concentration began, his respiration became slow, shallow, and rhythmical. Whereas GSR appeared occasionally before concentration, after commencement the line became comparatively flat. When he stopped sending his power, his breath again sped up, and the percipient's slowed down. His GSR showed excitement again, and hers calmed down.

When the percipient was asked how she had felt during the experiment, she said that all of a sudden she was aware of an oppressive energy, as though some enormous power were holding her down, and that she could not help but succumb to whatever it was that was affecting her.

The results of these agent-percipient tests suggest that it is possible for the mind to affect the functioning of another person's body without the use of any sensory agency. Mind is not merely a neurophysiological process; it is possible for the mind to transcend the physical world. The autonomic nervous system seems to be connected to the mechanism by which nonsensory levels of consciousness manifest.

The Greens recently performed a very similar experiment in which the famous healer Olga Worall acted as agent. Apparently, their results support my own. The more this type of strictly controlled experiment is done and the more corroborative evidence we can gather, the more clearly will this important message get through to the scientific community: *Consciousness has enormous potential. It can—and unconsciously does—link up directly with other minds and material forms. It need not be bound by the five senses, by time, nor by space.* As mystics have long held, consciousness is a unified totality; the divisive limits that we experience in everyday life are ultimately illusions.

CHAPTER 5

The Chakras

WE ARE now faced with evidence that consciousness can separate itself from the sensory dictates of the ordinary time-space continuum and directly affect the external environment. This fact demands an explanation, an expanded model of human nature that accounts for its existence.

Two such models, both of which afford a basis of theoretical explanation, already exist: the concept of chakra—nadi—subtle bodies, as developed most thoroughly in the Hindu culture (though references are made to the same perception in many other esoteric systems), and the Chinese system of subtle energies (which are manipulated in the medical practice of acupuncture). These two models were developed independently of each other, but they have many points in common and are mutually supportive.

Both models assume a much more complicated set of relationships between the physical and nonphysical aspects of our being than those which can be discovered through the western scientific method of investigation, relying heavily as it does on the dissection and examination of the physical entity alone. Further, and more importantly, both systems posit that human beings are manifestations of a universal continuum of energy and consciousness. The human organism is seen as being directly connected to the rest of the universe, both physically and psychically, through the continual exchange of energy and consciousness that takes place between them.

The next two chapters will provide a simplified description of these two theoretical models and describe some of the research I have done in an effort to examine them scientifically. Many terms I use may seem unfamiliar, and I realize that many people tend to discount evidence of new discoveries, no matter how convincing, if it doesn't fit into a preconceived framework. I have good reason to believe, however, that these models contain much valuable information, and that eventually a more concrete understanding of them will become an accepted part of the common store of knowledge. I expect that this understanding will completely revolutionize such fields as physics, medicine, and philosophy, bringing about a profound change in humanity's perception of reality—a change that has already begun.

Human physiology as perceived nonsensorily is quite distinct from what it is when seen by means of regular vision: there's much more to it than meets the eye. Nonsensory perception reveals that the human being is actually composed of three distinct bodies, not just the material one known to us. Esoteric teachings have termed these bodies the physical, the astral, and the causal. Each acts as a vehicle relating to a corresponding dimension of energy and consciousness. These bodies are seen to overlie and interpenetrate one another. The physical is the smallest of the three; the causal is the largest, extending beyond the physical body in a surrounding oval.

Just as the physical body is made of matter from the physical plane, the other two bodies are believed to consist of the "stuff" of their respective dimensions. A good introduction to the various planes on which a human being simultaneously manifests can be found in C. W. Leadbetter's *Man Visible and Invisible*, in which he describes the planes thus:

> In ordinary science we speak of an atom of oxygen, an atom of hydrogen, an atom of any of the substances which chemists call elements, the theory being that that is an element which cannot be further reduced, and that each of these elements has its atom—and an atom, as we may see from the Greek derivation of the word, means that which cannot be cut, or further subdivided. Occult science has always taught that all these so-called elements are not in the true sense of the word elements at all; that what we call an atom of oxygen or hydrogen can under certain circumstances be broken up. By repeating this breaking-up process it is found that there is one substance at the back of all the substances, and different combinations of its ultimate units give us what in chemistry are called atoms of oxygen or

hydrogen, gold or silver, lithium or platinum, etc. When these are all broken up we get back to a set of units which are all identical, except that some of them are positive and some negative.

The study of these units and of the possibilities of their combination is in itself one of most enthralling interest. Even these, however, are found to be units only from the points of view of our physical plane; that is to say, there are methods by which even they can be subdivided, but when they are so broken up they give us matter belonging to a different realm of nature. Yet this higher matter also is not simple but complex; and we find that it also exists in a series of states of its own, corresponding very fairly to the states of physical matter which we call solid, liquid, gaseous, or etheric. Again, by carrying on our process of subdivision far enough we reach another unit—the unit of that realm of nature to which occultists have given the name of the astral world.

Then the whole process may be repeated; for by further subdivision of that astral unit we find ourselves dealing with another still higher and more refined world, though a world which is still material. Once again we find matter existing in definitely marked conditions corresponding at that much higher level to the states with which we are familiar; and the result of our investigations brings us once again to a unit—the unit of this third great realm of nature [the causal]. So far as we know, there is no limit to this possibility of subdivision, but there is a very distinct limit to our capability of observing it. However, we can see enough to be certain of the existence of a considerable number of these different realms, each of which is in one sense a world in itself, though in another and wider sense all are parts of one stupendous whole.[1]

The astral body contains all desire and all emotions, and unconsciously determines much of our action when we exist within the supposed ordinary human limits. Both the physical and astral bodies are characterized by duality, the qualities of attraction and repulsion, positive and negative. The physical body sustains itself through the attraction and intake of air, food—all the elements that nourish and sustain it. When functioning in a healthy manner, it rejects all substances that are harmful or unnecessary to its maintenance, maintaining a balanced relationship to the external environment. When this balance is broken, sickness and abnormality result. Similarly, the astral body attracts what it desires and rejects what it does not desire.

Esoteric doctrines assert that between the astral and physical bodies is a connecting system of energy known as the etheric double, a state of energy which is considered to be physical but of a more subtle form of physicality than is presently known to science. (This type of energy appears to correspond to the *ki* of Chinese medicine, to be discussed in chapter 6). Modern technology is not yet sensitive enough to detect this form of energy, but many researchers now suspect its existence.

The causal body may be likened to the Christian concept of the spirit (as the astral body may be to the notion of soul), in that it is the highest part of our being, the part closest to "God." Yogic understanding posits a primordial Undifferentiated Oneness, an Absolute. At the moment of creation, the individual is differentiated from the One, and the causal body is born. The causal body, though individual, is identified with the Absolute; that is, it is beyond duality. (Accordingly, it is neither male nor female.) All the elements necessary for existence are combined here in the perfect equilibrium of pure life energy. The urge to realize this part of our selves and thereby to regain identity with the Absolute is the reason for all spiritual discipline.

Though it is beyond the distinction of subject/object duality, the causal body is individual and material, in the sense that it does have the power to manifest in diverse forms. The physical is the grossest form, the astral the subtler. When we are born, we are fully conscious only in the physical realm. As the evolution of consciousness progresses, one gradually becomes aware of and able to control the astral realm as well. As one becomes totally conscious of the ascending dimensions of being, one breaks through at last into awareness of the causal body, the springboard from which total reunification with absolute, undifferentiated consciousness can take place.

Imagine that the three bodies are concentric circles, with membranes in between the circles which connect them but prevent completely unrestricted interaction. We ordinarily experience existence primarily in the physical body, with a limited amount of energy and information passing through the membrane from the astral dimension to the physical, in the form of intuitions, dreams, memory, fantasy, et cetera. The membrane is necessary for an individual to function effectively in the physical realm; without it, conscious awareness would be flooded with information from the astral dimension. An unintended and violent disturbance of the membrane, such as may be induced through the ingestion of LSD, can induce psychosis. Awakening a chakra may be likened to a volitional and controlled thinning of the membrane which separates the various planes, making possible more communication between them.

The three bodies, it has been repeatedly claimed by yogic doctrine, are connected to one another through the seven chakras, which exist concurrently in the causal and astral dimensions, and whose function seems to reflect directly into the physical body. To nonsensory perception, chakras appear as "saucerlike depressions or vortices" in the surface of the etheric double. Leadbetter describes them: "When quite underdeveloped they appear as small circles about two inches in diameter, glowing dully in the ordinary man; but when awakened and vivified they are seen as blazing, coruscating whirlpools, much increased in size, and resembling miniature suns."[2] The two lower chakras are said to function primarily in the physical dimension, the two middle function primarily in the astral, and the three higher chakras represent ascending levels of the causal, or undifferentiated, plane. When a person abides in a state of sensory consciousness, the lower chakras work to maintain the functioning of the physical processes, while the higher operate at a minimal level in the physical dimension and remain "asleep" in their own dimensions. Spiritual evolution is seen to take place as each of these chakras is awakened to full knowledge of its own dimension.

Sensory consciousness, that which functions in the material dimension, tells us that we require certain physical substances, such as food and air, to exist. But, according to the yogic model, the aphorism that "Man does not live by bread alone" is a ridiculously gross understatement. Each level of our individual being is sustained by the energy from its own dimension, and the chakras are seen as the centers where this energy enters the various bodies. Some of the energy is distributed directly to the astral and causal bodies, which can accept it in nonphysical form; some is transmuted into lower-dimensional energy by the chakras and distributed where necessary. The ability that chakras allegedly have to transform the dimensionality of energy is a particularly important area of exploration, because proof that the chakras really can transform nonphysical energy into the physical would offer an important key to the real understanding of such diverse, hitherto inexplicable phenomena as psychic healing, teleportation, and the ability of certain yogis to live for sustained periods of time on no food and little or no oxygen.

Just as we are ordinarily unaware of the higher dimensions of our being, we are unaware of the constant exchange of energy and consciousness happening to us on these same levels. We are like a fetus, who receives continuous nourishment from its mother through the umbilical cord. Not until the child is born into this world as an autonomous individual does it begin to control the intake of sustenance to meet its needs. Similarly, once we awaken into the higher dimensions of ascending chakras, we can control energy at the corresponding level.

The chakras supply the energy that they accept from the outside world to the various bodies through a network of channels known as *nadi* in Sanskrit. There are fourteen main channels pervading the astral body, and thousands of tributaries. The three main channels are the *shushumna* (which runs along the spinal column, in the nonphysical dimensions) and the *ida* and *pingala* (lying to the left and right of the shushumna, respectively); the latter two transverse chakras and crisscross each other and the shushuma before finally terminating at the nostrils.

Nadi are said to have both a subtle and a physical form. Some modern researchers have initially assumed that the physical manifestation of nadis could be equated with the physical nerves. But deeper study, comparison of texts, and empirical evidence indicate that this is not the case. Instead, the "gross nadi" seem to be the exact same system known as the meridian system in Chinese medicine, though the two models were apparently developed independently of each other. For instance, it is written in the ancient Indian meridian text, the *Ayar Veda*, as well as in the classical acupuncture book attributed to the Yellow Emperor, that the nadi in the physical body and the acupuncture meridians are both filled with body liquid. It is also suggested in both sources that the energy involved exists as a layer between the purely physical and nonphysical aspects of our being. Such statements seem to cancel out the possibility that the gross nadi are nothing more than the physical nerves. My own theory is that the meridian system and the gross nadi are identical, and that they represent a physical but invisible system of physiological control and sustenance which is located throughout the body in the pervasive connecting tissue. I shall examine the evidence for the existence of this system and discuss the importance of its function in chapter 6.

The seven chakras are located along a central axis corresponding to the spinal column of the physical body. As reflected in the physical realm, they are each thought to control the physiological functioning of a certain biological system. In the astral realm each chakra is associated with a certain set of emotions, psychic tendencies, and nonsensory forms of perception/consciousness. In the causal realm, they are associated with increasingly unified experiences of reality.

Chakra-Plexus Relationship

Chakra	Plexus	Physiological System
Muladhara	Sacral and coccygeal	Genitourinary system connecting left and right sympathetic nervous system (corresponding to ida and pingala)
Swadhistana	Sacral	
Manipura	Solar	Digestive system
Anahata	Heart plexus in sympathetic trunk	Circulatory system
Visshuda	Cervical ganglia, connecting medulla oblongata with spinal cord	Respiratory system
Ajna	Hypophysis diencephalon	Autonomic nervous and and hormone system of the whole body
Sahasrara	Cerebral cortex	Nervous system, organs and tissues of the whole body

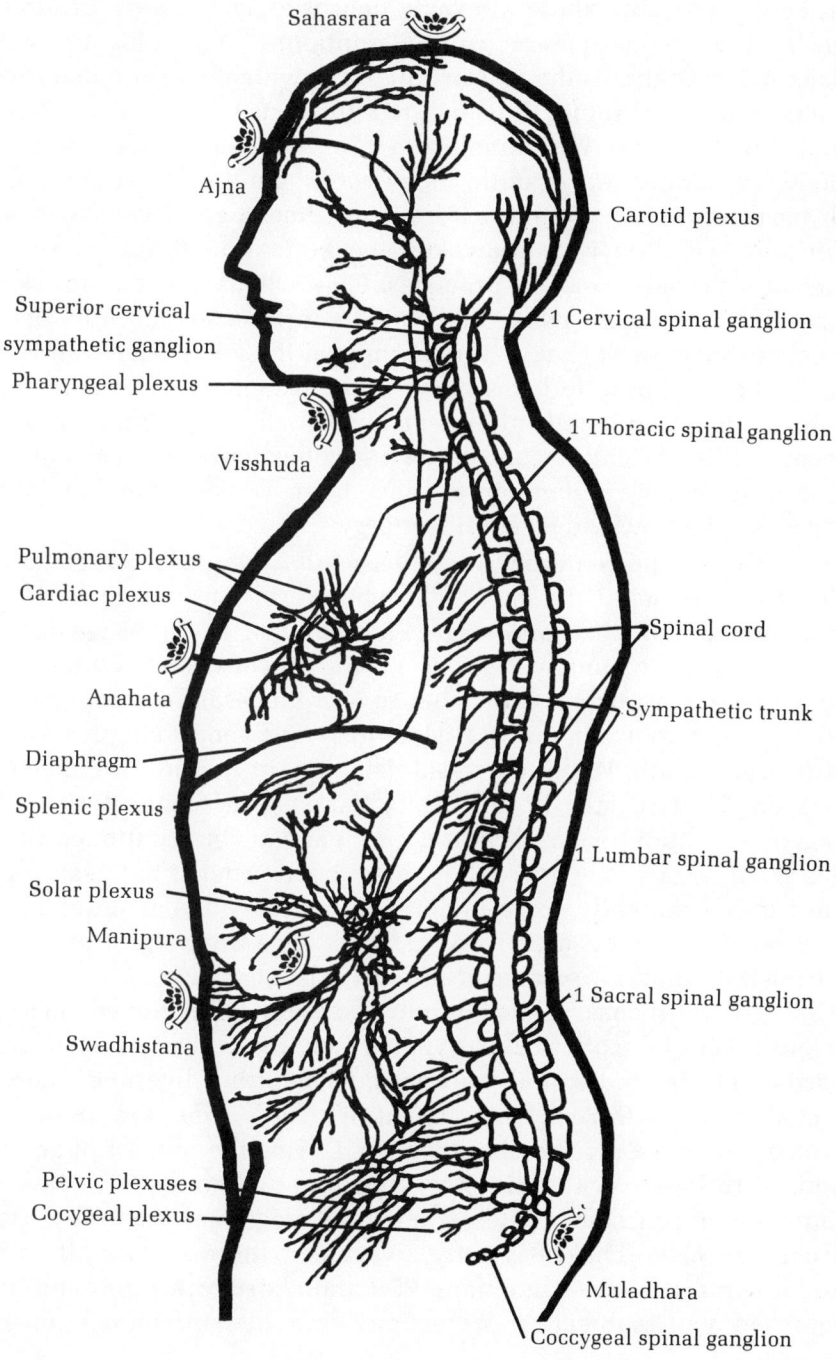

Figure 6. The seven chakras, or psychic centers, and the body's autonomic nerve plexuses.

The lowest of the chakras, the *muladhara*, sits at the base of the spine. The muladhara allegedly houses a very powerful form of energy known in Sanskrit as the *kundalini*. According to yogic teaching, kundalini is the a priori life force which in its original state is a part of the Absolute. It is the force that acts to mold each human being into his or her individual form, after which it dwells asleep on the higher planes while continuing to work to a limited extent in the physical dimension. One quick but sometimes dangerous system of spiritual evolution is to reawaken the kundalini through a set of specially designed ascetic practices. The activated kundalini rises through the chakras, awakening each as it proceeds, until at last it reaches the topmost chakra at the crown of the head, where unification with the Absolute takes place. Most systems of spiritual discipline do not intentionally attempt to awaken the kundalini, however, because it is difficult to control the enormous power released in the process unless the higher centers have been awakened and are able to control the flow and use of the energy.

The second center is called the *swadhistana;* it is located about 3 centimeters below the navel. This chakra governs the sexual functions and eliminatory processes. Physically, it is connected to the deeply rooted instinctive drives of the unconscious. When it is activated, the most noticeable change it produces is an abundance of energy, an increase in physical stamina, and a general improvement of overall health. While the swadhistana is in the process of being awakened, there may be a distinct increase in sexual desire and negative emotions such as hatred and jealousy, but continued practice purifies these feelings which have been brought to the surface, until they eventually come under a heretofore unknown degree of control. Vegetarianism and sexual abstinence are helpful practices during the initial awakening of the swadhistana center.

The third chakra, the *manipura* (spleen), is positioned near the solar plexus of the physical body. This center is said to govern a large portion of the individual unconscious and the digestive system. Included in its territory are the desire for power, the emotions, dreams, simple ESP, and clairvoyance. During the process of activation, disturbance of the digestive function, an enormous release of repressed unconscious contents, and great emotional instability often occur. Once the chakra is fully awakened, these factors all come under control. At the same time, ESP and clairvoyance become less confused with subjective projections from the individual unconscious, and therefore more objective and reliable.

The extrasensory perceptions received through the awakening of the manipura, however, are of a fairly limited, personal nature. In this stage of awareness individuals will be able to understand much more of the nature of reality than was possible before, but cannot do much to influence it. For example, they might now be able to see the astral body (aura) of another person and note some imbalance that may eventually manifest as physical or emotional sickness. But they do not yet possess the power to directly correct the astral imbalance, as one who has awakened higher centers could do.

The *anahata* (heart) chakra, located at the sternum, reputedly governs the heart and circulatory system, the beginning of a more far-reaching clairvoyance, and controlled psychokinetic power. This center, in its highly developed aspect, is the first that can truly be called spiritual; it is connected to forms of consciousness that transcend sensory-created distinctions and consequent egoistic desires—hence its association with universal love, with compassion free from vested interest. Awakening the anahata chakra brings a feeling of great expansion, the experience of moving beyond the tight shell of the small self. Higher mental faculties begin to come into play. Intuitive understanding of the real relationships among seemingly diverse elements of nature occurs, producing the type of intuitive leaps exhibited by artistic or scientific genius.

After this center has been awakened, if the lower centers are sufficiently purified (i.e., emptied of unconscious content), the experiencer will feel little attachment to the sensory or emotional realms. An ardent desire is born, however, that all beings develop to the point where they are equally as joyous in the realization of the interconnectedness and unity of all things.

The three higher chakras are the *visshuda* (at the throat), the *ajna* (between the eyebrows), and the *sahasrara* (at the crown of the head). Physically, the throat center is said to control respiration, the brow center to regulate the pineal and the pituitary glands; the sahasrara center, in the cerebral cortex, is considered the overseer and coordinator of the entire physiological system. To awaken the throat chakra one must abandon all desire and attachment to the material world; one may then function primarily in the causal realm.

The knowledge gained through the awakening of the three higher chakras surpasses the comprehension of the sensory-dependent intellect, as do the associated abilities. These are the levels at which true universal understanding takes place which many mystics hold to be the ultimate goal of the human adventure.

My primary interest in my research lay in determining whether any scientifically objective verification of the subtle body–chakra–nadi system could be found. Again, procedural problems abounded. It is not possible, using the tools of modern science, to investigate the direct relation between the physical and the nonphysical, because physical tools can only measure physical objects. At first all one can do is to attempt to clarify, through experimentation, whether there is a functional and positional relationship between the chakras and the physical body. The purpose of my research has not been to prove that the chakras exist in the physical body, but only to show that they do influence the function of the physiological system.

Clearly, the chakra-nadi system bears striking resemblances to the layout of the autonomic nervous system (as described in chapter 3). This similarity is especially remarkable in that the yogic model was developed through nonsensory perception centuries before the anatomical constituents of the nervous system were understood.

The sushumna seems to be in the exact same place as the spinal cord; the ida and pingala are similarly located along the main sympathetic nerves. Each chakra appears to have a corresponding nerve plexus (those bundles of nerve tissue which control respective areas of the body and the internal organs they contain), as designated in figure 6.

Assuming this correspondence, I decided to test an initial hypothesis: If alterations of consciousness are, as claimed by yogic teaching, inherently linked to the operation of the chakras, and the functioning of chakras tangibly affects the nervous system, then individuals who have allegedly experienced unification with higher states of consciousness should show some kind of corresponding alteration in the function of their nervous systems. The previous chapter presents the hypothesis that changes in consciousness do affect the autonomic nervous system, but no attempt was made to delineate specific chakra-plexus relationships. In this investigation, conducted over a fifteen-year period, I attempted to do just that.

The tests I used varied, but the basic objective remained the same: to investigate the workings of the organ system (e.g., stomach function) related to the nerve plexus (e.g., solar plexus) allegedly connected to a certain chakra (e.g., manipura), and to compare the findings obtained in terms of my original tripartite classification. That is, I took the test results of individuals who were subjectively and nonsensorily aware of activity in a certain chakra (specifically, members of group A and a few from group B), and compared these to those of the control, group C.

Several ways exist in which to investigate the condition of certain segments of the autonomic nervous system. I applied many methods in my experimentation on those parts of the system thought to be connected to chakras. All test results were subjected to strict statistical analysis. A partial summary of the experiments should suffice to demonstrate the findings and suggest some conclusions (details can be found in Appendix A).

In one of the early tests I conducted a detailed survey on the occurrence of disease tendency among 60 subjects (14 from group A, 11 from B, 35 from C). Subjects checked off physical symptoms they had experienced in relation to different organic functions (specifically those dealing with the circulatory system, respiratory system, digestive system, the skin, the genitourinary system, the ear-nose-oral cavity, the eye, and the anus), giving a mark for each classified symptom. The averages showed clearly that groups A and B far exceeded group C per symptoms known to be caused by autonomic nervous mediation (A = 264, B = 232, C = 59). This tendency was not seen among symptoms marginally connected to the autonomic system. Group A showed the highest degree of abnormality in the circulatory, digestive, and genitourinary functions (those functions supposedly connected to the chakras); B was second. Thus, those subjects who could be expected to show unusual activity in certain organic functions due to the activation of their chakras actually did manifest such abnormality. This finding supports the possibility of a connection between the autonomic nervous plexuses and the chakras.

In an attempt to discover whether any further physiological differences could be found among the three subject types, I did a long series of tests using a dermometer, a device which monitors the viscero-cutaneous reflex points (VCR). Modern physiology has revealed that nerve impulses generated by an abnormal condition in the internal organs are sent, by means of the autonomic nerves, through respective spinal segments to certain fixed points on the skin. This transmission of the impulse produces changes in the skin, muscle, minor arteries, and sweat glands surrounding the point, causing, in turn, alterations in electrical resistance and temperature, which may be read as an index of the abnormality.

By electrically stimulating selected VCR points and measuring the resultant skin resistance at a point on the palm, I began to notice some general tendencies. Control subjects showed little difference in skin current values before and after stimulation; this tells us that the autonomic nervous function of the internal organs was stable, that the stimulus was acceptable to the system. In the case of many subjects from groups A and B, however, I found a significant difference between pre- and post-stimulation states (see appendix B). Often this difference would show up more markedly in a specific organ, and the same three organs—heart, stomach, and kidney—were found to be the most unstable. These are the three organs controlled by the plexuses with hypothetical chakra connections. From the fact that an increase of current at the palm reflex points may generally be

Figure 7.   Disease tendency among groups A, B, and C.

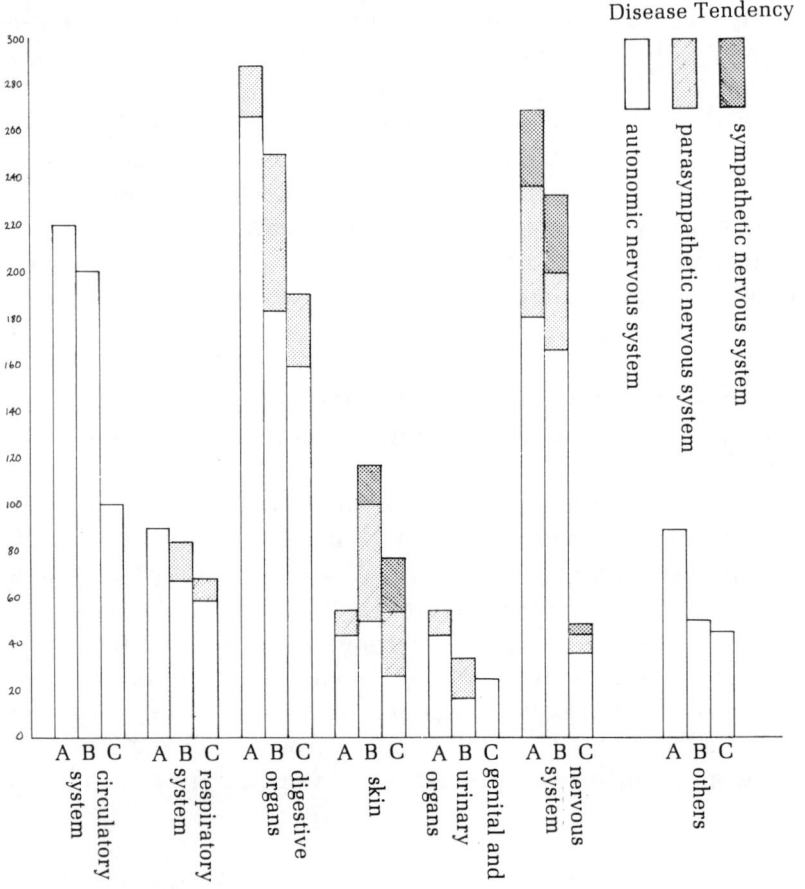

interpreted to indicate excitement and strain of sympathetic nerves, whereas a decrease of the current value would mean that the reaction of parasympathetic nerves predominated, I deduced that in group A subjects each organ's reaction to stimulation was predominantly parasympathetic, whereas in group B sympathetic nervous reaction predominated; control subjects, on the other hand, manifested no autonomic instability or imbalance between sympathetic and parasympathetic nervous components.

Blood pressure measurements taken before and after a stimulus was applied to the left VCR point for the heart, examination of dislocations of spinal vertebrae, and polygraph readings revealed similar differences in the autonomic functioning of the three groups.

These studies all suggested a correspondence between the nonsensorily perceived chakras and the nerve plexuses, but the methods of testing were still rather indirect. A more direct approach would be to determine whether a subject could alter the bioelectric potential being emitted from the plexuses through mental concentration on the chakras. This hypothesis was derived from the yogic assertion that the chakras themselves are energy transformers. Since all cells emit some electrical charge, I suspected that the chakras might be found to effect singularly significant changes in the energy being emitted from the body, especially in those individuals who claim to have awakened their chakras. (This is not to suggest that the energy being emitted from the chakras is itself physical; only that emission of higher-dimensional energy might cause reverberations in the physical dimension.) Therefore, I invented a machine that attempts, in an original manner, to measure the energy being ejected from the body.

I first designed an electrode that can measure the body's electrical charge without touching the skin. For all present methods of bioelectric measurement, electrodes must be in direct contact with the skin surface. This contact produces a potential, the magnitude of which (some hundreds of microvolts to some tens of millivolts) is much larger than that of the ordinary electrical potential produced by the body. Therefore, in most types of bioelectric measurement, only those components of the total bioelectrical signal which vary over time are detected, and the time-invariant or slowly varying components, which may contain hitherto-unknown information, are simply ignored.

This problem led me to devise a machine that is capable of

Figure 8. Diagrammatic representation of the device used to measure the energy ejected from the body through the chakras.

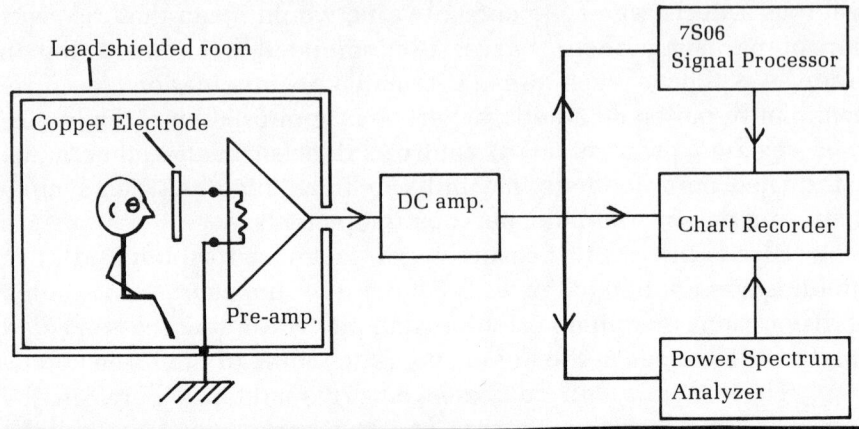

Figure 9. Data of an awakened and controlled chakra.

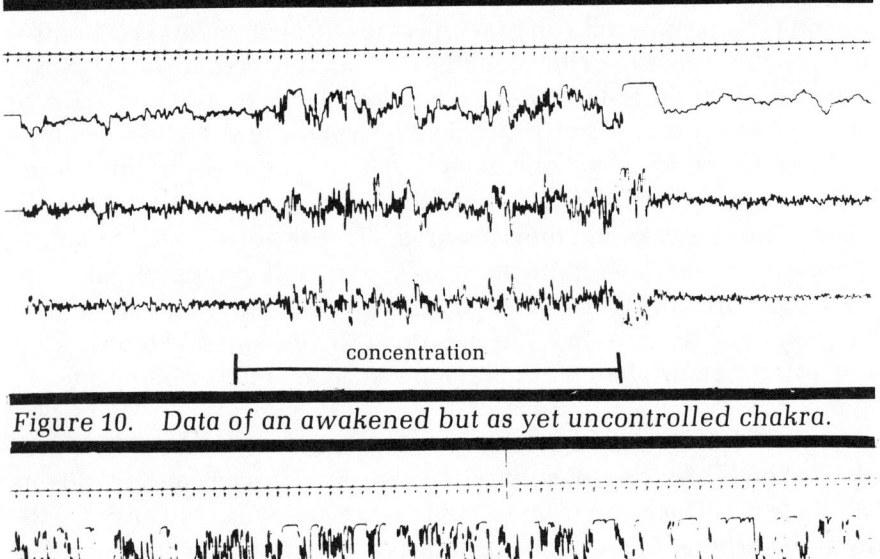

Figure 10. Data of an awakened but as yet uncontrolled chakra.

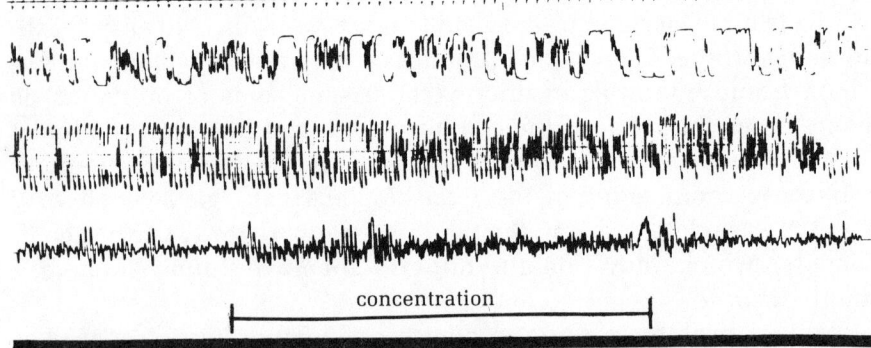

measuring the potential of the electric field surrounding the subject's body without the need to attach any electrodes. The machine, which is placed in a specially designed light-tight, lead-shielded room, uses a movable copper electrode with a sinusoidal oscillation; the electrode can be positioned in front of any part of the subject's body. The intensity as well as the fluctuation of the electric field potential around the subject's body is detected through the capacitive coupling between the electrode, the subject, and the room. The signal thus generated is first passed through an extremely high input-impedance (800 megaohm) preamplifier, which then sends the measurement information via a set of DC-amplifiers to a signal processor and a power spectrum analyzer outside the room, where the data is recorded on a standard chart recorder.

An early test done with a prototype of the machine yielded some interesting data. An electrode was placed in front of various parts of the subject's body in turn. The recording mechanism was started; the subject was instructed to remain quiet for 30 seconds, then to concentrate or to try to eject energy from that part of the body in front of which the electrodes were placed (30 seconds), and then to stop concentrating and again to remain quiet for 30 seconds. When the electrode was placed in front of a chakra the subject claimed had been awakened, the potential and frequency of the electric field asssociated with energy ejection from the body were significantly greater than those found in control subjects.

The data reproduced in figure 9 are those taken on one of my yoga students who had been practicing meditation for over five years and claimed to be able to control energy ejection from his ajna chakra to some extent. The intensity of the signal shows marked differences before, during, and after concentration. This difference suggests to me that his chakra is awakened and that he is able to control it. The student claims—and my perception of his aura concurs—that his swadhistana chakra is also active, but that he is not able to control the ejection of energy from it. When the same test was performed placing the electrode before the lower chakra, the results reflected this distinction (see figure 10). The great amplitude of the graph would be accounted for by the fact that he has awakened this chakra to some extent; that there is not much difference between the periods before, during, and after concentration would suggest that, if so, he has not yet come to control it.

In this early experiment, the electrode was not completely

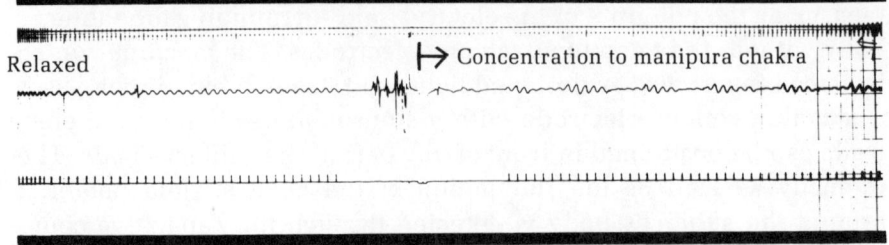

Figure 11. Data of an awakened, controlled manipura chakra.

fixed; it vibrated only slightly, but enough to prevent exact measurement. Moreover, the electric field inside the experiment room was not sufficiently eliminated, which also interfered with exact measurement of the body's electric field. Though this disturbance was not great enough to invalidate the findings of this experiment, I decided to make further measurements even more exact by improving the electrodes and their supporters so that they did not vibrate, and by covering the room with aluminum.

Figure 11 shows sample data taken with the improved machine. The subject had been meditating for five years and perceived that her manipura chakra had become activated. For a year or so before the data were taken, the subject evidenced many of the symptoms that I have come to associate with group B types and particularly the awakening of the manipura chakra: emotional instability, stomach trouble, sudden changes in vitality level, and an increase in ESP experiences, particularly clairvoyance. I decided to run the following test using the improved machine. The electrode was set in front of the midpoint of the solar plexus (the supposed location of the manipura). The subject was instructed to concentrate on the chakra and to try to eject energy through it. Remarkably, during the time that she felt she was able to project energy, the electrical pattern being recorded by the apparatus disappeared. In order to ensure that this change was not due to her respiration pattern, we had her breathe normally and then hold her breath as long as possible, without concentrating on the chakra; no change was produced. The subject was then asked to concentrate on another chakra (the ajna), one which she had not concentrated on during meditation practice. No change in the electric field was recorded, and the subject did not signal that she subjectively felt activation of the chakra. The dramatic change produced in the electric field surrounding the subject when she concentrated on her solar plexus seems to be directly related to the activation of the manipura chakra, an activation of which the subject herself was conscious.

Another set of data were taken on a subject whose condition suggested exceptionally strong but as yet uncontrolled working of the anahata chakra, evidenced in a marked psychokinetic ability. Her mother is a psychic who has precognitive ability. The subject also had many precognitive experiences from the time she was a young girl. When she was fifteen, she had a prophetic vision about President Kennedy's assassination, which occurred a few weeks later. A few years ago, she came to Japan and here met someone with whom she discovered a strong psychic affinity, someone previously unaware of such phenomena; telepathy between them began to occur spontaneously, and they worked to develop this ability. After both attended a demonstration by the well-known psychic Uri Geller, PK phenomena began to manifest in their presence: clocks and watches would often stop or become erratic, keys and spoons would bend inexplicably. Such phenomena were frequently observed while the subjects were at the institute for testing.

The subject herself had said that she was very aware of some kind of energy phenomenon occurring in the area of her heart. Therefore, I decided to place the electrode in front of her heart; I instructed her to concentrate on the anahata chakra and to try to eject energy through it. She was told to push a certain button when she felt energy being ejected, in order to alert the experimenters in the next room. We noticed the continual ejection of high-frequency energy from the heart area, but no pulsation was evident.

Besides the usual electrode, a photo-electric cell for measuring light (the electrode room is completely dark during the experiment) was also set in front of the chakra involved. During the experiment, a signal was registered in the channel of the photo-electric cell implying that light was somehow created. How can this remarkable phenomenon be explained? I do not know. Yet, somehow, when the subject concentrated on her anahata chakra, light was produced.

Moreover, a number of apparently PK phenomena occurred during this experiment. The metal pen of the EEG machine, which was not in use during the experiment, bent out of shape, and both the electric clock on the wall and the experimenter's watch stopped exactly at 3:15, the time at which the subject intended to leave for another appointment. Were these phenomena in any way related to the fact that the subject was concentrating on her anahata chakra? Perhaps.

The results of the chakra machine tests strongly suggest that a nonphysical cause (mental concentration on the empirically known chakras) can produce direct physical results. These findings, which require further corroboration, merit further investigation, because they support the yogic assertions that chakras act as bridges between the various dimensions of existence. More thorough research into the chakras may provide invaluable information about the relationship between consciousness and matter.

Chapter 6

Ki and the Chinese Meridian System

A SUBTLE system of life-sustaining energy that circulates throughout the physical body was postulated 4500 years ago in China. This energy, called ki (pronounced "key"), is characterized as the "vital force" upon which gross physical life is dependent.

Scientific researchers have recently begun fairly extensive investigation into the ancient Chinese claims. As early as 1937, Sir Thomas Lewis stated in a paper entitled "An Unknown Nervous System" (published in the *British Medical Journal*), that he had found "a cutaneous nervous system 'unsuspected at present,' which is unrelated to the known sensory-nerve pathways and unconnected with the sympathetic nervous system." Modern bioelectric pioneers have made similar discoveries. Marshal Gilula, M.D., and James Beal, an electrical engineer, comment:

> Recent reports... have shown the existence of another system which functions to transmit information and control signals on or near the surface of the body. This system, which developed prior to the nerve systems in evolution, is involved with basic functions of growth, pain sensations, wound healing, regeneration of organs and limbs, biological rhythms and responses to changes in the environment (weather, time-of-day, location). The nature of this system makes it easily influenced by internal and external electrical conditions. It appears to connect with the nervous system (through the Schwann cells which enfold the nerve fibers) at particular locations which can be easily found by their different electrical properties."[1]

I am mainly interested in this system as a potential link between consciousness and matter. Before discussing this problem, however, I should like first to offer some historical background on the Chinese explanation of this system and then describe some of the physiological research that supports the empirical assertions of its existence.

Many mystics have intuitively perceived some form of energy basic to the maintenance of the body: it has no obvious physical form, but is perceived as a "force" underlying physical manifestation. Early Chinese cosmology made note of this energy, and experiments with it over thousands of years of clinical practice produced oriental medicine.

According to the ancient Chinese cosmology, the original, undifferentiated Absolute divided itself into two aspects, yin and yang. This doctrine embodies the basic tenet of the mystical view: All existence is basically unified; dualistic attributes (such as positive and negative) are merely manifestations of the essential Absolute.

The Chinese stress the complementarity and relativity of polar opposites. Positive can only exist as long as negative exists; positive taken to its extreme becomes negative, and vice versa. This duality implies no moralism. It is, rather, a characterization of the two fundamental life forces which create and sustain existence, and an expression of the implicit harmony that exists between them. All of nature maintains a balance between positive and negative. Only when this balance is upset, does disharmony, abnormality, or "evil" occur.

In Chinese cosmology, the human body is seen to be a microcosm of the universe, regulated by the same laws, the same principles that govern the interaction of the positive and negative forces. This ancient hypothesis is evident in every aspect of physiological activity: in the polarity between the sympathetic and parasympathetic nervous systems, in the ionization (into positive and negative charges) of all metabolic activity, in the expansion and contraction of blood vessels. But, Chinese medicine claims, even more basic to the existence of the physiological organism than biochemical processes is the interaction of the yin and yang aspects of ki energy.

The Chinese emphasis on ki energy resulted from the discovery of certain physiological phenomena which gave rise to an understanding of body function that differs from the contemporary western model. One concise account of this development, found in *The Layman's Guide to Acupuncture,* runs as follows: "In some manner the ancient Chinese became aware of an increased sensitivity of certain skin areas (called points) when a body organ or function was impaired. It was observed that in all patients the same skin areas became hypersensitive in the presence of a specific illness or organic dysfunction. Moreover, the sensitive areas varied consistently according to the organ function deviating from the norm. It was thus that some of the relationships among various internal organs and their functions were observed and established."[2]

Gradually it was recognized that these points follow an identical mapping in all individuals, and that the points connected to one type of organic function fall along a continuous line. These pathways of energy flow became known as *ching,* or meridians.

The Chinese hypothesized that ki has both positive and negative aspects, the interaction of which regulates the body's physiological activities. When there is an imbalance in the function of ki, physical or mental illness will result. The object of acupuncture is to restore the body's balance, either by stimulating the energy flow into an area that is insufficient in ki or by dispersing an excess of the energy accumulated at specific points located along a meridian.

The main meridians discovered by the Chinese appear to correspond to the gross nadi perceived nonsensorily by the Indian mystics. Two different cultures, two different modes of perception (one physical and one not) both attest to the same system of energy in the body. I do not claim that either the nadi or the meridian system is a final, definitive description of the nonneurological control network—only that their very formulation suggests the existence of an important part of the human body, a part about which far too little is known.

I am also aware that this brief introduction to acupuncture is necessarily an oversimplification; however, I am not concerned here with building a case for acupuncture (though the long history of continued acupuncture practice and the modern research underway all strongly suggest its effectiveness). I am interested, rather, in a number of conjectures that emerge from even a cursory study of the subject. First, it seems that acupuncture meridians and the physical nervous system do not coincide positionally; they are independent of each other. The Chinese still cannot explain why acupuncture works; they only know it does. A clarification of the seemingly nonneural mechanism responsible for the effectiveness of acupuncture should contribute to a greater understanding of the human organism. Also, if, as the mystics claim, the subtle energy system acts as a mediator between mind and body, a more thorough investigation into ki may help to clarify the way in which consciousness and the body are connected.

I myself came to the study of ki and the corresponding acupuncture methodology in a rather abrupt manner. As the result of years of meditation practice, I had become directly aware of a basic vital energy circulating throughout my subtle and physical bodies. This I associated mainly with the nonphysical nadi and chakras of Indian teaching, and I did not overly concern myself with its direct physical manifestation; instead, I studied the potential relationships between the chakra-nadi and the nervous systems. But personal experience suddenly led me to consider the nonneural yet physiological aspect of the system I had perceived.

During the sixties, my spiritual guide, Odaisama, became ill, and I was compelled to assume many duties at the shrine. These activities were added to a heavy research schedule. One of my main duties at the shrine was therapeutic counseling, for which I intentionally attempted to activate my manipura chakra in order to gain paranormal knowledge about the deeper causes of people's physical and emotional problems. After three years of relying exclusively on the manipura for this type of information, I began to notice quite a strain on my system. I chose to ignore the warning signals, however, because, except for some ear trouble as a child, I had never suffered any illness.

Around this time, I was invited to lecture on the psychophysiological concomitants of altered states of consciousness at the Andhra University in India. The position meant a very demanding schedule; in addition, I was using any spare time to conduct physiological investigations on yoga practitioners. I ended up by developing a severe stomach problem. At the same time a bad case of eczema appeared along the outside of both legs, from knee to heel. It didn't occur to me to connect the stomach and the leg until I noticed that the rash was running exactly along the stomach meridian as designated by Chinese acupuncture theory.

Figure 12. Stomach meridian, leg portion shown. Energy flows along this meridian from the head to the foot.

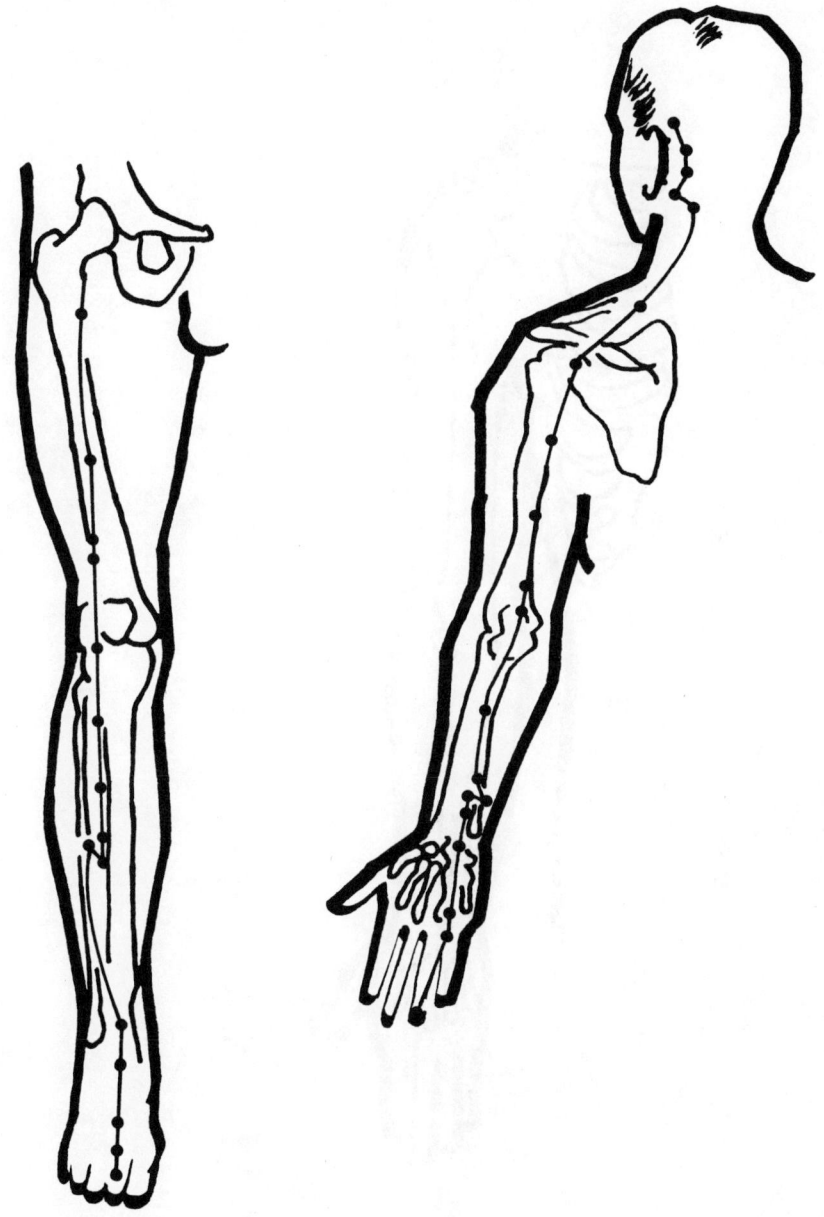

Figure 13. The triple heater meridian. Energy flows along this meridian from the hand to the head.

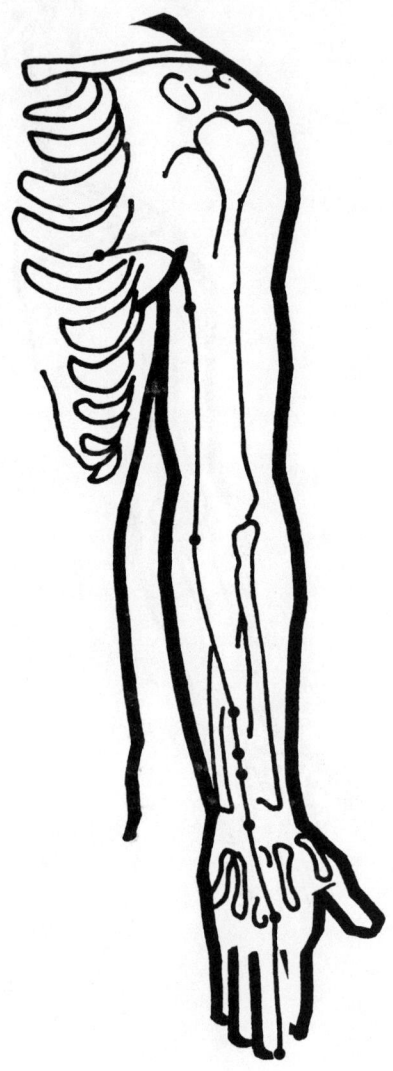

Figure 14. The heart constrictor meridian.

Curious, I decided to investigate the matter further upon my return to Japan. I first had an X-ray taken, which in fact revealed the presence of a stomach ulcer. The stomach has a number of viscerocutaneous reflex points (VCR) on the abdomen and back. These points become noticeably sensitive if a stomach disorder occurs. I checked, and noted marked sensitivity at these points. When I consulted acupuncture texts on the subject, I was surprised to find that these were also acupuncture points directly connected to the function of the stomach (the *chukan, ryumon, iyu,* and *hiyu*). I was intrigued by this correspondence—another instance of ancient knowledge confirming the modern. But I found even more interesting the fact that the neurophysiological model of VCR points cannot account for the leg rash, nor is there any direct nervous connection between the stomach and the exact portion of the leg where the rash occurred. Yet such a relationship is assumed by meridian theory.

My experience suggested to me that perhaps the ulcer reflected a functional abnormality of the ki in the stomach meridian (supported by the evidence of the leg rash), which in turn was induced by overstimulation of the manipura chakra. I conjectured that ki might be a mediator between the nonphysical and the physical aspects of the human organism. A whole world of relationships opened up to me, connecting chakras, the subtle body, nonsensory consciousness, ki, and the physical body/consciousness. I decided to study acupuncture in depth.

After having obtained detailed knowledge of the acupuncture system, I began a series of experiments to try to investigate scientifically the existence of the meridians. Oriental medicine holds that three main types of ki energy work together to sustain the existence of the physical body, from birth to death: upper energy functions in the heart and lungs, and is physically introduced into the body through oxygen; the middle energy functions in the digestive organs and is introduced through food; the lower energy acts in the genitourinary system and is introduced through the absorption of nourishment.

Acupuncture theory and practice delineate twelve major meridians, through which ki flows. Ten correspond to organs: the lungs, large intestine, stomach, spleen-pancreas, heart, small intestine, bladder, kidneys, gall bladder, and liver. These meridians are identified with more than just the actual organ, however; they are related to the whole system of which, in the western model, the organ is considered to be the center. (The heart meridian, for example, is responsible for the entire process of circulation—from the action of the heart muscle to the distribution of oxygen in the blood.)

Two meridians with no known correspondence in western medicine are the triple heater and the heart constrictor. Both are concerned with the function of the whole organism, rather than any specific organ. The triple heater is thought to control the energy (ki) level of the whole body. The triple heater meridian is the path through which the three types of ki flow throughout the body; it controls their distribution. There is no corresponding organ in physical anatomy, but the reputed location of the intangible triple heater is the same as that of the swadhistana chakra, the center which the Indian mystics held to be in control of the physiological plane.

The heart constrictor meridian exists in an antagonistic yet complementary (yin/yang) relation to the triple heater meridian, and functions together with it to control the energy level of the entire organism. The heart constrictor meridian is also intimately connected to the function of the heart.

The network of meridians is symmetrical and begins at the triple heater meridian point in the center of the belly. The triple heater meridian runs from here to the lungs, where it becomes the lung meridian, and, as such, extends along each arm to the tip of the thumb. A branch connects the lung meridian with the large intestine meridian at the tip of the index finger, and this large intestine meridian runs up along the opposite side of the arm, then branches downwards to the abdominal area and upward to the jaw and nasal area, where it connects with the stomach meridian. In this way, the various meridians connect with one another, forming a continuous circuit of energy along the hands, feet, and torso until finally the twelfth meridian (the liver meridian) connects with the first meridian (the lung meridian) in the lungs, completing the circle.

The meridians are thought to be paired. The lung and large intestine meridians, for example, are so connected; this means that an excess of energy in one meridian is usually accompanied by a corresponding lack of energy in the other. (A blockage of energy in the lung meridian due to common cold, for instance, will generally manifest as a lack of energy in the large intestine, giving rise to problems such as diarrhea.)

These twelve major meridians are related to a number of "extraordinary" meridians. Two of these extraordinary meridians are the conception vessel meridian and the governor vessel meridian. Both run parallel to the spinal column (the conception in front, the governor in back). Positive (yang) energy is supposed to center along the governor meridian, and negative (yin) energy along the conception meridian.

Traditional acupuncturists use various methods of examination to diagnose the functional condition of meridians. Many utilize the sense of touch and recognition of various pulses located throughout the body. Some practitioners have sufficient sensitivity actually to feel the subtle ki itself; they are thus able to detect immediately any imbalance or abnormality. To understand the intricate relationships that exist among the various meridians requires years of experience in both diagnosis and treatment.

Each meridian has specific points that most accurately reflect different aspects of its functional condition. Every point along the meridian is capable of reflecting organic disturbance to some extent, but two sets of points not located on the meridians themselves are particularly effective for monitoring the condition of the meridians. The "alarm" point, located along the frontal midline of the torso, reacts whenever the balance of ki in a given meridian is disturbed; the "associated" point, located along the spine, reflects abnormality in the activity of the associated organ.

It must be clearly noted that neurologically, the meridians are not directly connected to the organs themselves or to the alarm and associated points; *physical nerves do not run between them*. I chose to investigate the reputed relationship between the meridians and their alarm-associated points, because if such a relationship was demonstrable, it would serve as evidence that the body has features that are unaccountable in terms of the present western physical model, and that the Chinese account may be more complete. In other words, before attempting any complex research concerning the meridian system, I wished first to gather more evidence that the meridians indeed exist in a physically apprehensible framework.

Though the type of energy circulating throughout the meridians is not yet clearly understood, clinical practice is based on the fact that this energy can be influenced by physical manipulation. In modern times, it has been demonstrated repeatedly that the acupuncture points have a different degree of electric resistance than the surrounding areas of skin and that the electrical measurement changes in response to the manipulation of points along a meridian

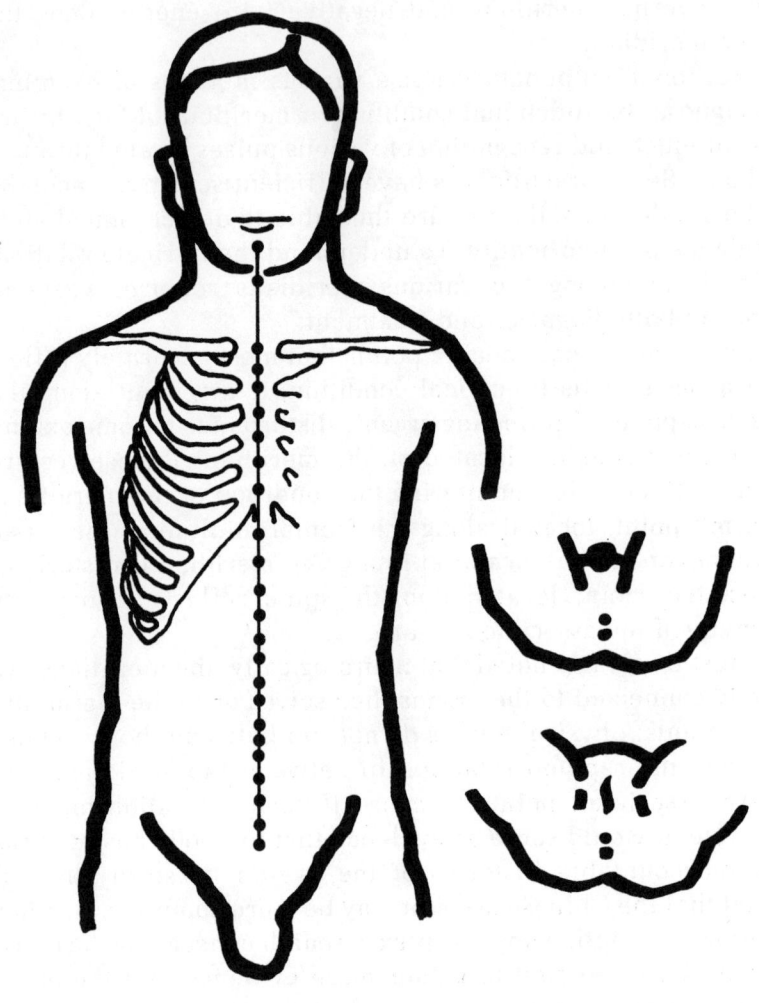

Figure 15. The conception vessel meridian.

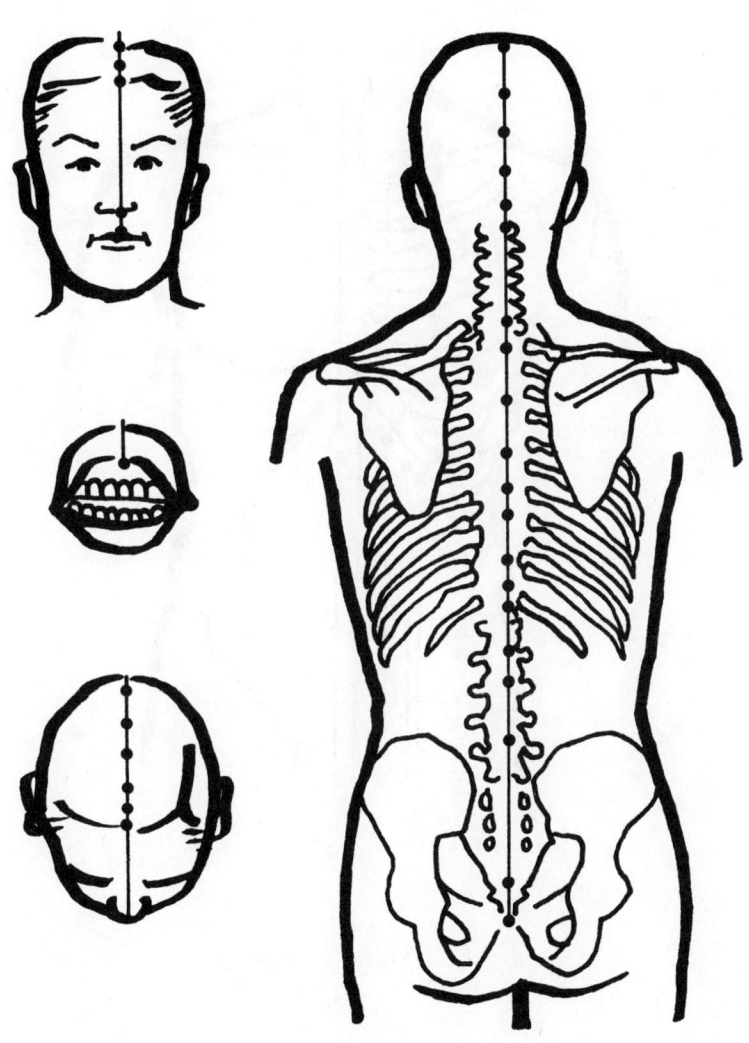

Figure 16.   The governor vessel meridian.

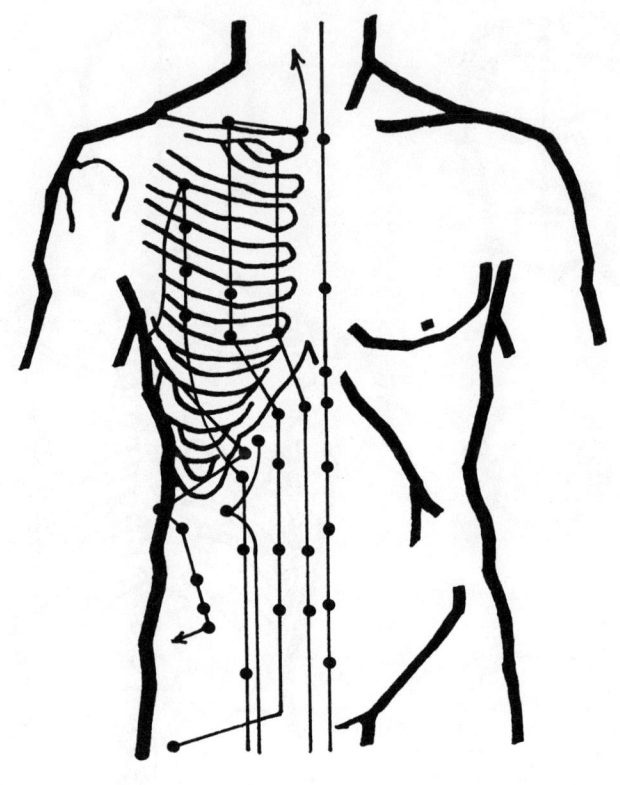

Figure 17. The alarm points.

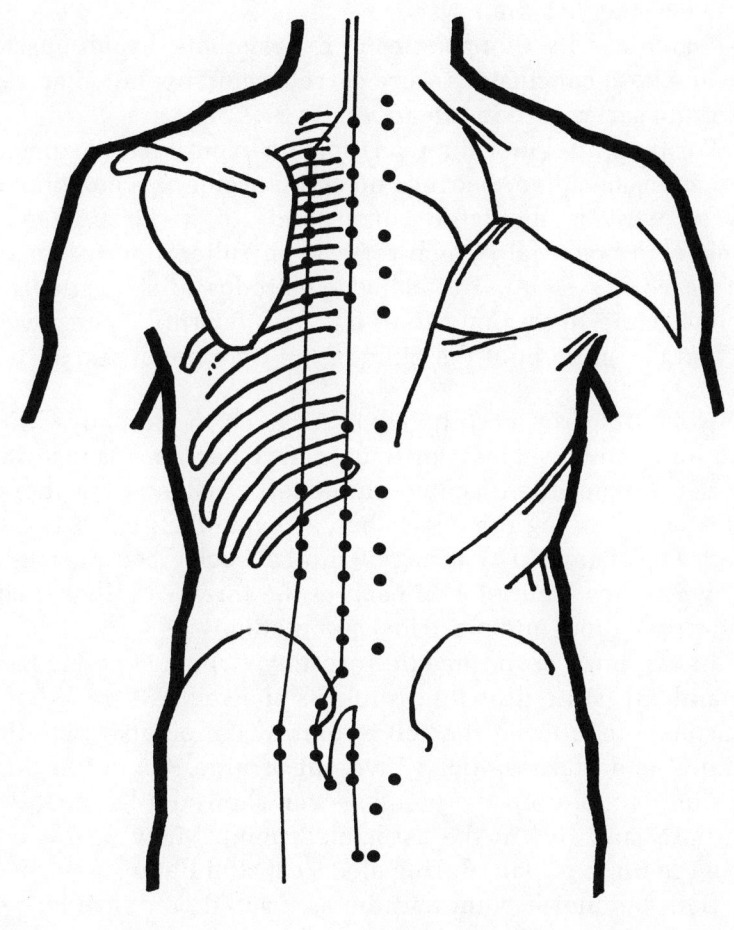

Figure 18. The associated points.

(through, for example, the insertion of acupuncture needles). This evidence led me to form the hypothesis that if one point of a meridian were stimulated (electrically or otherwise) and a subsequent electronic response were recorded in other points along the meridian, as well as in the alarm and associated points (but not in other places), such findings would constitute strong evidence corroborating the existence of the meridian system.

I conducted six long series of experiments. I shall describe in part two whose conclusions were corroborated by the others; a summary of the series is found in appendix B.

For the initial meridian verification experiment, I selected the triple heater meridian, because no organ or nervous complex recognized by western medicine corresponds to it. First, to measure galvanic skin potential (which reflects any alteration in the sympathetic nervous system), I attached electrodes at six points located along the triple heater meridian on the left arm, at the associated point and alarm point at the palm of the right hand, and at the right wrist.

A control test—with no stimulus given to the body—was conducted for 2 minutes. Next, an acupuncture needle was inserted into a point of the meridian located on the left wrist, and left there for 2 minutes before being removed; then a slight electrical stimulus was applied. The change in electric potential at each electrode was measured by EEG for 2 minutes in each of the three conditions: control, needle stimulation, and electrical stimulation.

In response to the needle insertion (which subjects reported was painless), six out of nine subjects showed GSP reaction at all measuring points; three reacted at four of the points (including the associated and alarm points). Five out of nine reacted to the electrical stimulation at all the points; seven showed a large GSP at the alarm point and five at the associated point. Many of the subjects reported feelings of pain during electrical stimulation.

Both the alarm point and the associated point are located too far from the wrist to be easily affected by a change in electric field around the source point; however, the change found at these points was in fact *larger* than the changes registered at points on the meridian located closer to the point of stimulation. Since there is no neurological connection between the point stimulated and the point exhibiting a definite reaction to the stimulus, another system of physiological communication is indicated. It is also interesting that these findings back up the experiential knowledge of acupuncture practice, which holds that the alarm and associated points, though not located on the regular meridian, are nevertheless affected by the condition of the meridian.

In a similar experiment, I applied electrical stimulation to the end point of the triple heater meridian, located on the tip of the fourth finger, to see whether a change in galvanic skin potential occurred at any of the aforesaid points. I found that when a strong electrical stimulus was applied to the end point, a jump in GSP was noted at all the points and at the right palm. However, when a weaker stimulus was given GSP occurred *only* in the associated and alarm points. I also did this test in reverse, stimulating the associated point, and found significant GSP in the end point at the finger.

Again, the alarm and associated points are located too far from the point of stimulation for one to assume that the change in GSP found in them was caused by a change in the electromagnetic field surrounding the stimulation point. Moreover, there is no known nervous connection between the tip of the fourth finger and the corresponding associated point or alarm point. Therefore, these tests provided challenging evidence that the meridian system does exist.

Other researchers are doing various experiments aimed at verifying the existence of meridians. (Dr. Yoshio Nakatani in Osaka and Professor Masayoshi Hyodo of the Osaka Medical University Pain Clinic have done important work in this field, using a bioelectric investigation of acupuncture points in relation to syndromes designated in western medicine.) I find particularly valuable the work of Dr. Yoshio Nagahama,[3] a physician associated with the medical school of Chiba University. He has done significant research into the correlation between the results of diagnosis and treatment based on both eastern and western medicine. During his research, Dr. Nagahama came across a patient who had been struck by lightning and could clearly feel the "echo" or movement of current produced by an acupuncture needle. Dr. Nagahama investigated this echo by inserting a needle in the source point of each meridian and having the patient trace with his finger the course along which the echo was transmitted, while describing the echo's relative strength. Dr. Nagahama used a stopwatch to measure the time it took the echo to travel from the source point to the opposite terminal of the meridian. The patient himself was not familiar with acupuncture theory, but the courses he traced followed all the traditional meridian pathways. Dr. Nagahama further found that the transmission speed of the needle echo was very much slower than that of a neural impulse in the spinal or autonomic nerves; for needle echo, the transmission speed was 15.2—48.1 cm/sec, whereas impulses in the spinal or autonomic nerves travel much more quickly at a speed of 5—80 m/sec. This evidence offers further support to the hypothesis that the meridians constitute a nonneural system.

Nagahama carried out another investigation with this patient to explore the special relationship said to exist among the meridian, its associated point, and its alarm point. When he inserted a needle in each of these points, he found that an echo from the needle did reach the related meridian.

These experiments, along with those cited in appendix B, plus the findings of other researchers, gave me confidence that the system of subtle energy of which I had become aware during meditation practice does exist physiologically. I needed to find a reliable and consistent way of measuring the system in order to incorporate it into my research methodology.

An experience involving my wife helped lead me in the right direction. We have five children, and my wife leads a very busy life. Occasionally she becomes over-tired, and during such times I could feel a noticeable difference in the temperatures of her left and right fourth fingertips. I measured the temperature of these points when my wife was feeling tired and found a difference of 4—6° C. I then tested the galvanic skin potential and found a great difference here as well.

Now, the tip of the fourth finger is the initial point of the triple heater meridian (the meridian supposedly in control of the general energy level of the body). I checked further into traditional acupuncture texts and discovered that a time-tested method of diagnosis is to check the fingertip points of the meridians for coldness and insensitivity to heat; this practice is based on the clinical observation that the fingertip points of a meridian whose corresponding organ is diseased will evidence such symptoms.

Taking a clue from my wife's condition I decided to experiment with a large number of subjects. Fourteen meridians have terminal points (*seiketsu*) at the tips of fingers and toes. These are the twelve meridians already mentioned, plus the diaphragm meridian

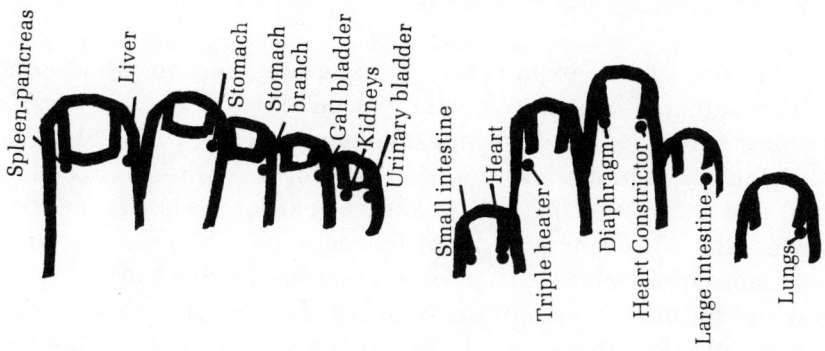

Figure 19. The terminal points (seiketsu) of the meridians on the fingers and toes.

(related to respiration and to the level of nervous excitation in the body) and the stomach branch meridian (connected to digestion). After attaching electrodes to the finger and toe end points of these fourteen meridians (*seiketsu*) I measured both skin resistance and temperature at all twenty-eight points. Then I calculated the difference in measured values between left- and right-side end points. Where great differences were found, indicating imbalance in the meridians, other methods of diagnosis were applied for corroboration. One of the immediate findings was that many of those subjects who complained of lack of energy and lethargy, and who had consulted doctors to no avail, manifested a great difference in the values recorded at the fingertip points of the triple heater meridian.

Though each meridian contains a number of points that clearly reflect the activity of its ki (points of supply, accumulation, release, intersection, etc.), the finger and toe end points of each meridian seem to be particularly well suited for monitoring the condition of the rest of the meridian and the function of the corresponding organ.[4] Not only can they be located quickly, easily, and accurately, but they reflect the condition of certain deep-seated organs (such as the kidney, liver, and spleen) that are difficult, costly, and time-consuming to diagnose with existing western methodology. I hypothesized that bioelectrical measurement of the seiketsu points might prove to be a reliable method of diagnosing organic pathology. Moreover, if the underlying theory is correct, evidence of abnormality in the meridian activity should appear before a physical ailment. The ability to measure abnormality, instability, excitation, or weakness of meridian function before the manifestation of organic disease would prove a great boon in terms of preventive medicine.

To test these hypotheses, I experimented on about one hundred subjects over a two-year period. First, I measured skin resistance values by attaching negative electrodes at all the bilateral points (the points of the meridians on the fingers of the left and right hands) and a positive electrode to another part of the body, and then recorded the differences. The differences in the skin resistance values measured before and after stimulation were recorded at all initial and terminal points. If the energy is flowing smoothly along a meridian, the difference in skin resistance values between the left and right seiketsu points should be slight. A great difference would indicate an imbalance in the activity of the meridian and a disturbance in the corresponding organ.

I was excited to find a clear corroboration between the results of seiketsu measurement and more orthodox methods of diagnosis. For the majority of healthy subjects the deviation was indeed small: about 2.0 (range: 0.5–5). Values outside this range did appear to indicate abnormality. One subject, for example, showed a left-side value of 5.5 in the heart meridian in contrast to a right-side value of 28, giving an extremely high difference. Since the difference indicated severe heart disturbance, I asked the subject to have an ECG examination, which showed over-excitement of the heart ventricle. Thus, a dangerous situation was clearly detected through the meridian-based method of diagnosis.

I asked the entire group of subjects to fill out questionnaires concerning their personal disease history and tendencies. I found a statistically significant correlation between current differences in the left/right end points and the disease histories, suggesting that seiketsu measurement is able to detect chronic as well as acute ailments.

An important focus of the research was to compare the results of the meridian testing with medical diagnoses made independently by physicians trained in standard medicine. I selected subjects who showed a high degree of difference and obtained reports from their doctors; conversely, I measured the differences of subjects with known medical syndromes to determine whether a corresponding meridian imbalance could be found. Finally, having amassed a large body of corroborative data, I was satisfied that a valid relationship between the bioelectric measurement of meridian endpoints and medical diagnoses indeed exists. My records report many cases in support of this theory; I shall relate a few to indicate the type of correspondence found.

An early case involved a man brought to the institute by his daughter for acupuncture diagnosis. During a routine medical checkup, he had been diagnosed by X-ray as having stomach cancer, although he did not show any obvious symptoms. As is customary in Japan, the doctor did not tell him about the cancer; instead, he was told that he was developing a stomach ulcer and would have to undergo appropriate treatment. The family was told the truth, and the daughter told me.

When I examined this patient, using the seiketsu method, I discovered a very large difference between the left and right stomach meridian points, indicating the seriousness of the situation. But the triple heater meridian points showed no great difference, which may explain why the patient still had a lot of energy and felt fine. (I have seen this pattern repeatedly; in many cases, it is not until the triple heater meridian shows dangerous imbalance that physical symptoms become noticeable.)

In another early case one of my female assistants was acting as a research subject. She felt healthy, but I discovered large differences in her bladder and kidney values. About a week later she began having trouble with urination, visited a doctor, and learned she had a severe case of cystitis.

The positive results of the first two years of seiketsu measurement encouraged me to try to quantify the measurements further. Years of research went into developing an efficient machine and data standard capable of translating terminal point measurement into information applicable to the western diagnostic framework. The first task was to create an apparatus that could measure and compare the meridian functions of large numbers of people—one that could be used by other researchers, acupuncturists, and medical professionals. The second task was to develop a standard by which this form of measurement could be compared to other, more widely known forms of diagnosis.

I named the machine I developed the "Apparatus for Measuring the Function of the Meridians and Corresponding Internal Organs"—AMI for short. This computerized apparatus can diagnose imbalance anywhere in the physiological organism in a matter of minutes. The machine's twenty-eight electrodes are first attached to the twenty-eight meridian end points on the fingers and toes, and then 3 volts DC charge is applied sequentially through the electrodes in order to measure the skin currents at respective finger and toe tips.

Figure 20. The AMI Machine: the Apparatus for Measuring the Function of the Meridians and Corresponding Internal Organs.

The machine records the initial skin current as well as the equilibrium skin current (measured 1 second later). The initial skin current, denoted by BP ("before polarization"), reflects the body's basic constitutional state or metabolic level, whereas the equilibrium skin current indicates the temporary (i.e., acute rather than chronic) condition of the meridian concerned. The difference between the two (BP - AP) is defined as P ("polarization"), which indicates the amount of resistance to the external environment which the body is able to manifest.

    The AMI is superior to standard GSR devices, which are designed to measure only AP values, in that it can measure BP and P values. This is a distinct advantage, since AP values are subject to alteration under the influence of temperature and the mental and physical condition of subjects, whereas BP and P values are relatively constant and can tell us more about the chronic state of the organism.[5]

    After measuring over five thousand subjects with this machine and comparing their data with the results of conventional medical examinations and their reported symptoms, I was able to draw up criteria of normality and abnormality for the values recorded BP, P, and AP. These standards are programmed into a computer. Patients' raw data are fed into the computer, which calculates their status in relation to the criteria for normality; any abnormality is signaled in the printout.

I feel the AMI may prove useful in a number of different ways. My hope is that enough hospitals and clinics will collect sufficient data with this machine so that comprehensive criteria for known medical syndromes can be outlined.

Already the Kanagawa Prefectural Rehabilitation Center is carrying out investigations to compare the results of biochemical and X-ray analyses with the diagnostic findings of the AMI, similar to the verification investigations I conducted at the Institute.

The Bob Hope Parkinson Research Institute in Miami has used the AMI to determine whether any characteristic patterns of electrical energy are found in patients with Parkinson's disease. Researchers Marshall Gilula and James Beal feel the machine may be able to define both known medical problems and problems not yet recognized by conventional diagnosis, as well as the effects of medication on the physiological system. They hope thus to be able to distinguish different stages of types of Parkinsonism and perhaps to gain information that could help patients control the disease process. My institute has collaborated by making statistical analyses of their raw AMI data; these reveal considerable metabolic and bilateral imbalances in the bodies of Parkinsonism patients and abnormalities in the heart, small intestine, and large intestine meridians analogous to cases of rheumatism and arthritis. In this way we hope to correlate AMI findings with disease patterns.

One prefectural government in Japan is using the AMI to screen all employees during the annual required medical examinations (only those with abnormalities must undergo further tests). Eventually, the AMI may preclude the need for expensive, painful, and time-consuming diagnostic tests. Once this machine has discovered a malfunction, such biochemical tests may describe the condition in greater detail, but the likelihood of their being conducted unnecessarily—as is so often the case—will be greatly reduced. At the same time, the AMI can detect disease tendencies, so it can catch some diseases before they blossom into pathological conditions requiring intensive medical care.

This chapter has dealt with the ki/meridian system in its physical aspects. Let us now turn to a discussion of ki's larger implications vis-à-vis the nonphysical dimensions of existence, chakras, and altered states of consciousness.

Chapter 7

Actualizing Human Potential

MYSTICAL TRADITIONS assert that human beings are interconnected and mutually involved in a gradual process whose aim is ultimate realization of the absolute, non-dualistic reality. Many sources have taught that the human being is not alone, not separate from the universe, but, in fact, essentially unified with all being. Sense-dependent consciousness belies this truth, suggesting that we lead an unconnected—perhaps meaningless—existence. Science is beginning to investigate the mystical, empirical claims.

Through my research, I have attempted to demonstrate concretely that the human being is more than just a body and a limited intellect. The steps that I and others have taken in exploring the connection between the physical and nonphysical sides of being may be small but are nonetheless powerful in that they lead inevitably toward the destruction of the limitations that prevent actualization of the human potential.

A model of human nature which includes the chakras and ki may be the prototype of an eventual clarification of the relationship of consciousness to matter, as well as the relationship of consciousness to the rest of existence. I have used the tools that I have developed (the AMI and chakra machine), as well as conventional devices, to explore the implications and applications of the chakra/meridian system in relation to both ordinary and nonsensory consciousness.

Through long-term research on the three subject types, I have come to some very interesting generalizations, which are becoming clearer as the work proceeds.

As I mentioned earlier, I determined that the range of competitive activity between the sympathetic and parasympathetic nervous systems in the psychically developed individual is much wider than is normal, though the system as a whole remains dynamically balanced. I therefore hypothesized that this phenomenon is attributable to the inherent connection between the chakras and the autonomic nervous system.

Now, assuming that the meridian system is even more closely related to the nonphysical aspects of our being than are the gross physical constituents of our nature (as empirical evidence suggests), we should expect the development of consciousness to produce a distinct change in the functioning of the meridian system. Using this assumption as a working hypothesis, I used the AMI machine to examine over five thousand people (according to the tripartite classification), and the following specifics emerged: When a certain chakra has been awakened, the meridians or nadi that receive energy from that particular chakra show abnormalities in their activities and functions (notably, an excessive flow of energy). At the same time, when a subject willfully activates a certain chakra and realizes the attendant nonsensory states of consciousness, or if the subject possesses spontaneous but uncontrolled psychic abilities, connected meridians exhibit instability.

In particular, I found two psychic types which appear to be quite common and which are easily determined by the AMI machine: the manipura-type and the anahata-type.

The sort of ESP encountered in the majority of people showing psychic abilities is traditionally thought to be related to the manipura chakra (the center able to receive energy directly from the outside as information on the astral level). I have found repeatedly, in tests with the AMI machine, that those who have demonstrable ESP ability show abnormality in the spleen and stomach meridians, the two meridians most closely related in function to the manipura chakra as described by yogic theory. The manipura chakra is also thought to govern the subconscious, the emotions, and the digestive system. When subjects appear to be in the process of awakening the manipura chakra, in addition to showing abnormalities in the spleen and stomach meridians they evidence emotional instability and digestive trouble, even though no corresponding pathological (organic) disturbances can be found by other methods of diagnosis.

On the other hand, individuals who have PK ability (for instance, psychic healers) consistently display abnormalities in the meridian of the kidney and bladder, and, most frequently, the heart constrictor meridian. Those who cannot control their PK ability, but in whom PK phenomena manifest spontaneously, tend to show the abnormalities only in the kidney and bladder meridians. Yogic theory holds that controlled PK ability is associated with the heart chakra and uncontrolled ability with the swadhistana chakra (corresponding to the kidney and bladder). My findings seem to corroborate this empirical knowledge.

A description of a few of the case histories encountered in the research should help explicate the findings. In chapter 5, I provided chakra machine data about a subject who was suspected of having awakened his swadhistana chakra. The intense discipline he was undergoing at the time left him somewhat low in energy, a characteristic I have noted in connection with many group B subjects. Before the chakra machine experiment was done, I measured this subject with the AMI machine and found that his triple heater meridian showed a definite lack of energy. I then measured the electrical potential emanating from the alarm and associated points of the triple heater meridian. (The associated point is located directly over the place where the swadhistana chakra is alleged to be.)

I found that little low-frequency energy was being emitted from the alarm point in front, whereas much was being discharged from the associated point in back. The frequency of the waves from the front was in the fast, 100–1800-cycle range. On the other hand, the potential of the energy from the back was very high (around 100 millivolts), but the waves were very slow (1–80 Herz). According to the AMI investigation, the subject's lung meridian showed a slight excess of energy, and when the lung meridian's alarm and associated points were measured as a control, it was found that these parts were emitting high-frequency energy (200–800 Herz) and a very low potential (under 10 millivolts). The frequency of energy ejected from the back points was higher and the potential lower in this lung meridian than in the triple heater meridian; moreover, the same tendency was observed in those meridians which showed no abnormality according to the AMI investigation. This corroboration is especially significant in view of the fact that the triple heater meridian measurement was taken by the AMI at the tips of the fourth fingers—points not connected neurologically to the alarm or associated points.

We can infer from these findings that when a meridian shows a lack of energy, high potential, low-frequency energy is likely to be emitted from the associated point of the meridian, points which positionally correspond to chakras. In the case of this particular subject, the data, viewed as a whole, indicate a definite relation between meridians, chakras, and paranormal abilities, as well as between states of consciousness and states of health. Subjectively, this individual, who had been performing a severe regimen of meditative practice aimed at awakening the swadhistana chakra, always felt tired and was prone to experience spontaneous psychokinetic phenomena. Objectively, AMI investigation showed that his triple heater meridian was depleted in energy. Further, when the alarm point of the triple heater meridian—corresponding to the swadhistana chakra—was measured by the chakra machine, I found relatively low frequency energy of high potential being emitted as compared to the alarm point of the lung meridian which appeared normal in the same AMI diagnosis. I have found similar correspondences in many subjects who are in the midst of practicing some form of spiritual discipline.

Another subject described earlier, who apparently was able to nullify the electric field around her by concentrating on the manipura chakra, had her AMI data taken a few minutes before the chakra machine test was done. The stomach and spleen meridian, whose corresponding organs are those supposedly controlled by the manipura chakra, show definite abnormality. The subject reported no discomfort in her spleen or stomach areas and suffered none subsequently; these abnormalities seem to indicate vivification of the manipura chakra and related meridian function rather than a pathological condition.

In yet another subject, the one who seemed to produce light by concentrating on her anahata chakra, abnormalities were found in the stomach, spleen, and heart constrictor meridians in concurrent AMI investigations. This suggests to me that both her manipura and anahata chakras are somewhat active.

I have found repeatedly that people with PK ability evidence symptoms that seem to indicate active anahata chakras. For example, I once tested a Catholic priest, Father Francis McNutt, who is gaining a worldwide reputation for mass healings produced through his prayers. Father McNutt professes that the healings take place solely through the grace of Jesus Christ and that he is simply a channel through which the grace flows. I examined him with the AMI, and his data provided a striking example of the anahata type under discussion: the heart and heart constrictor meridians evidenced distinct abnormalities (i.e., excess of energy). This excess indicates to me that Father McNutt's anahata chakra is probably awakened and that the associated PK (in this case, healing ability), is of a developed nature.

We have encountered many similar examples. Once a psychic came to visit me at the institute, complaining that before a phenomenon such as a spoon bending occurred, her heart would begin thumping wildly. She had noticed the connection but had no idea what it meant. I recorded her AMI data, which showed great instability in the heart function. The heart thumping and the AMI data together suggest that her anahata chakra is active (reflected in the instability which manifests as an excess of energy) but not yet awakened.

Other types of scientific evidence point to the connection between the heart and PK ability. One well-documented case is that of a Russian woman, Madame Kulagina, who has exhibited on film—as well as to a large number of scientists—an extraordinary degree of psychokinetic control. She is able to suspend a ping-pong ball in midair and to move objects across a table top without touching them. She was subjected to detailed electrophysiological tests, and it was found that her pulse rises to as high as 240 during such feats. After suffering a heart attack, Kulagina had to stop acting as a research subject. Clearly, her case offers further support for the thesis that PK, the heart chakra, and the heart activity are related.

While attending the International Psychotronics Conference in Tokyo in spring 1977, I witnessed a demonstration by the famous French psychic, Jean-Pierre Gerard, during which he moved a number of objects across a table top without using any visible physical means. To bear out my thesis, I suggested that his pulse be taken while he was executing the allegedly psychokinetic phenomena: I predicted that it would rise. Many witnesses were astonished to find that a leap in his pulse rate did occur, but it really is not surprising at all, considering the possibility of an anahata-heart-PK connection.

The majority of individuals with highly developed psychokinetic ability seem to employ it for purposes of psychic healing—perhaps as a result of the compassion that is said to accompany the awakening of the anahata chakra.

I have done extensive tests on different types of healers, and it seems to me that many of them, consciously or unconsciously, perform the healing by tapping into higher dimensional energy through the heart chakra and sending it into the patient, to correct the energy imbalance behind the disease manifestation. When the anahata chakra is activated, it appears to function as the agent of transformation, as the entity capable of changing the dimensionality of energy. Experimental results, plus close observation of healers at work, have led me to suspect that when the anahata chakra is awakened, the energy released is often radiated through the end point of the heart constrictor meridian, located at the tip of the third finger, or through the palm of the hand, which lies directly in the path of this meridian.

This transformation of the dimensionality of energy is understandable in light of the three-body theory, which states that the human being has a physical, astral, and causal body. The energy of the causal body is said to be impervious to imbalance, because it is not composed of positive and negative aspects. The astral body, on the other hand, is dualistic by nature; therefore, it is liable to become imbalanced. Such instability usually manifests as mental and emotional disturbance. This disturbance, if unchecked, will be reflected in the body, resulting in physical illness.

It seems that the ki flowing throughout the meridians is the primary agency which introduces the astral disturbance into the physical body, and that ki is the type of energy that most healers (e.g., laying on of hands, acupressure) inject into the body of a patient to bring about cures. Empirical evidence has led me to believe that it is not that difficult to become conscious of ki and to learn to manipulate it. A more sophisticated type of healing, however, is to tap the energy of the causal body, through activation of the anahata chakra or even higher chakras, and introduce this exceedingly powerful energy into the suffering individual (through, for example, long-distance healing). People who have the ability to use the energy of the causal dimension in this manner can effect difficult mental and emotional cures, because causal energy is stronger than either ki or astral energy.

In an attempt to demonstrate the mechanism of introducing nonphysical energy into patients in order to stabilize their condition, I performed the following experiment. This particular subject had been practicing a strict program of meditation for three years and was showing signs of the breakthrough into nonsensory states of consciousness, a stage commonly accompanied by general instability.

AMI electrodes were attached to the subject's fingertips and toes. Measurements were taken in a five-run control. After a short interval, another control of five runs was taken, and the average value of each control set was computed. I was thus able to determine the extent of change occurring in the subject attributable solely to the passage of time. A test was performed to determine statistical significance between the two average values, and a significant difference was found in the meridians of the lungs, large intestine, diaphragm, heart, and small intestine. Since the subject was simply sitting quietly, the significant changes indicated definite instability in physiological function.

Since I wished the subject not to be aware of the nature of the experiment or of the fact that I would be sending him energy, I was not able to use the nadi/meridian system and eject ki energy through my palm or finger, an action which would have been obvious. I attempted, rather, to send energy directly through my anahata chakra, and I subjectively felt that I had made contact and entered him.

We took AMI measurements five times before I began to direct energy toward him and then repeated them throughout the period of transmission. All of the meridians that had varied became stabilized, except for the diaphragm and heart constrictor, which continued to display a slight change. Thus, apparently it is possible to stabilize ki imbalances in the body of another individual if one has learned to control the ejection of energy from one's own body.

One type of healing that uses mentally directed, nonsensory energy to produce changes in a patient is psychic surgery, a subject that has interested me for many years. This type of healing is most highly developed in the Philippines, though instances have been reported elsewhere (for example, in Brazil). Though clearly a certain percentage of the claims made by psychic surgeons are definitely fraudulent, I have gathered evidence that some psychic surgeons are actually able to remove diseased tissue from a patient without using any sensory means.

I have had independent laboratories perform bio-chemical tests on a number of tissue and blood specimens taken during operations I witnessed in which I saw no evidence of a physical instrument being used by the surgeon. No visible scars remained on the patients' bodies, so it is difficult to imagine how blood or tissue could have been removed by any means other than psychokinetic action.

In one of these analyses, done in November 1972, I took

samples of blood released from five subjects' bodies during the operations. Each subject also allowed me to take a blood sample directly from their lumbar area, after the operation. The surgeon, Tony Agpoa, was informed of my intentions only a few minutes before I took the samples. Upon my return to Japan, I sent both sets of specimens to a university hospital to check whether the two specimens from each patient did in fact belong to the same person. In each case the blood types matched.

At the Annual Convention of the International Association of Religion and Parapsychology in Tokyo in 1977, Agpoa repeatedly demonstrated his ability to cut through thick wads of adhesive tape using the endpoint of his small finger. He claims that the energy that he ejects from the endpoint of his small finger to cut the tape is the same sort of energy he uses to make incisions in his patients. As most psychic healing is "bloodless," I asked Tony why he finds it necessary to produce immediate, physically apprehensible results (such as blood) during the operation. Tony said that it is actually unnecessary to produce blood or tissue to cure a person, but he held that people will become more convinced of the existence of the higher dimensions of being if they witness the materialization of physical objects through nonphysical means, and that such immediate experience is necessary to shock materialistically oriented people into spiritual growth.

Most healers do not go to such lengths, but simply inject healing energy into patients to effect a cure. Most healers attempt to go to the heart of the problem by reestablishing balance within the astral body and allowing physical rehabilitation to follow naturally. The fact that certain individuals are able to control the transference of nonsensory, higher-dimensional energy to such a degree suggests that all of us probably unconsciously interact with one another, in a number of ways. On the physical dimension we may seem separate from one another, so that interaction is only by "choice," but on ascending dimensions (those known to the mystics but not visible or immediately obvious to sensory intellect), we are increasingly interrelated.

I decided to use the AMI apparatus to investigate the way in which we unconsciously influence one another; I suspected that the ki-meridian system is more subtle (less material) than the physical being and therefore more closely related to the nonphysical dimensions of existence. In a series of single-pair tests, I randomly designated two pairs of subjects and, for control purposes, measured the two separately a number of times. Right after the control runs, I put the two subjects together in the same room and measured them simultaneously. The subjects remained silent. Although no communication appeared to be taking place, statistically significant differences were found in the workings of the meridian system. These differences were due, I hypothesized, solely to the fact that the subjects were near each other, and that their respective ki systems were influencing each other. Continued repetition of the experiment yielded similar results.

This mutual influence which people in spacial proximity exert—whether consciously or unconsciously—on one another's meridian system seems to be a subtle psychokinetic effect. Such phenomena make one wonder just how deeply we affect each other in everyday situations without realizing it.

The next series of this experiment followed the same design. However, I chose, rather than random subjects, pairs who appeared to have intrinsic active/passive relationships (e.g., mother-child, father-child, older sibling-younger sibling). After subjecting the test results to statistical analysis, I found that the subjects assumed to be psychologically "receptive" showed a significantly greater degree of change in their data than did the partners suspected to be more active. I noted, moreover, that the meridians most affected were those which had already shown some abnormality in the independent control runs. This suggests that a weak point in the body's organic systems is particularly susceptible to the influence of other people.

Similarly, when I tested combinations of subjects from the presumed groups (A, B, C) I discovered that those subjects presumed to be more developed psychically (group A) affected group B individuals more strongly than they did group C subjects. It would seem that the group B subjects, already having slight cracks in the walls of their sensorily determined beings, are more open to other forms of energy than are those subjects more tightly enveloped by the physical realm. The results of this experiment led me to believe that psychokinetic ability is most effective when both agent and percipient are somewhat psychic.

In the A-B-C experimental series, the specific meridians showing the greatest changes were those of the kidney and the heart, and one branch of the stomach meridian concerned with the organic function of the diaphragm (which corresponds to the solar plexus). The triple heater and urinary bladder meridians were also found to be affected. Correlations of the data with the first series of the single-pair tests indicate that the kidney and the triple heater meridians seem to be most susceptible to receiving some sort of energy transmitted (consciously or unconsciously) by another person.

Chinese medicine has traditionally considered the triple heater to be the organ that regulates the interaction of the three divisions of vital energy and the flow of energy throughout the body. Chinese medicine also posits an inborn life force, said to dwell in the kidneys; life is thought to be maintained through the cooperation of the kidney and triple heater. Thus, it is interesting that these meridians should be so susceptible to the influence of others.

The results of these AMI tests indicate that the energy flowing in and out of the meridians can interact quite fluidly with that of other human beings. Learning, through a basically mental process, to control the ejection of this energy or of even higher forms of energy appears to be the key to understanding many heretofore inexplicable phenomena such as psychokinesis and psychic healing—phenomena that most likely depend on this energy for their manifestation.

The research I have been doing and the devices I am developing are applicable to another area of endeavor, that of understanding the changes which consciousness undergoes as it evolves and attempting to avoid certain difficulties which may be encountered during the process.

Awakening to nonsensory states of consciousness is a difficult process, apt to throw the individual into periods of great confusion. This confusion often manifests as mental, emotional, and/or physical instability. If the instability is extreme, someone undergoing such an awakening may appear critically ill or psychotic. Just as birth into the physical plane is often a traumatic experience, awakening to other dimensions of being may be extremely disturbing.

The seemingly pathological symptoms that are apt to appear during the process of spiritual evolution are signs of growth, not of sickness. This fact is not generally known by those not versed in the mystical traditions. At this time, with so many people hungering for actualization of their potential beyond sensory limitations, it seems clear that we must develop ways in which to distinguish between pathology and the changes that occur naturally during the process of spiritual breakthrough. Such criteria will help seekers by letting them know what to expect and by putting their experiences into perspective, as well as aid professionals (medical doctors, psychiatrists, psychologists) who might be called on for a diagnosis.

The necessity of objectifying the symptoms concomitant to spiritual growth is engendering an important, albeit small, movement to distinguish between "breakdown" and "breakthrough." Some psychologists, for example, now believe that in certain cases, "schizophrenia" may be a perfectly normal stage in the organic process of the evolution of consciousness. R.D. Laing notes that:

> "Schizophrenia is itself a natural way of healing our own appalling state of alienation called normality . . . Madness need not be all breakdown . . . It may also be breakthrough. It is potentially liberation and renewal as well as enslavement and existential death. It is not an illness to be treated, but a "voyage." Socially, madness may be a form in which . . . often through quite ordinary people, the light begins to break through the cracks in our all-too-closed minds."[1]

Claudio Naranjo says:

> The danger of psychosis that besets the legendary sorcerer's apprentice is today a matter of great interest, because we are beginning to see that not only is psychosis the outcome of a failure of the ego (to deal with the unconscious) but also a state of potentialities greater than those of the normal states. Julian Silverman has remarked on how a shaman undergoes, as a part of his initiation process, something that we would diagnose as a psychotic state. He is not hospitalized for it and "treated," but, quite to the contrary, his state is respected and allowed to follow its natural course. The consequent question is then: are not some of the syndromes that we treat as schizophrenic, tumultous, and even cataclysmic, stages of development that we are, for lack of trust, interrupting instead of allowing them to take a positive course?[2]

Perhaps we can understand the reason for the potentially

traumatic effects of spiritual breakthrough better by examining the problem in light of the changes that occur in the mind through the meditative discipline.

The "normal" state of consciousness is one in which conscious awareness and the related individual unconscious function together, in dynamic balance. Neither is completely independent, yet each sphere possesses a degree of independence. The conscious self is clearly somewhat independent: we possess a "will" which enables us to act in accordance with our beliefs and views on life (often overriding contradictory feelings and desires). Conversely, the strength of an unconscious drive often will assert its independence by undermining any conscious attempt to suppress it. We see the emergence of the unconscious frequently when the conscious self has been weakened through exhaustion, alcohol, or a violent emotion, and unconscious forces suddenly gain control. Dream and fantasy also attest to the ability of the unconscious to interrupt the normal function of waking consciousness and assert a degree of independence. Sometimes, an element of the individual unconscious may be so strong that it takes complete possession of consciousness—as in delusions and hallucinations.

The interaction of these two regions, conscious and unconscious, produces the character and behavior of the individual. In the normal state, the conscious operation of reason or understanding is dominant, while semiconscious or unconscious desires and feelings are suppressed. The normal individual projects a unified personality and mode of action.

Spiritual discipline tampers with this balance. The prerequisite of meditation and subsequent unification with other states of consciousness is that the normal functioning of consciousness be suspended, so that the suppressed energy of the individual unconscious may be released and the meditator may break through the limits of sense-created consciousness into nonsensory states.

The object of the meditative discipline is to understand the individual unconsciousness, and to transcend it. This objective is not reached directly or easily. The individual unconscious has tremendous power to create perceptions of reality that seem real but are not based on objective fact. Defusing the power of the unconscious, so that one can perceive sensory and nonsensory existence objectively, is an arduous task.

Hypnosis demonstrates just how much power the unconscious

possesses to create its own reality. The hypnotic state is one in which the normal function of waking consciousness has been stopped and the individual unconscious is free to enter the realm of awareness (moreover, the individual has, to some extent, relinquished the power of volition—which is not the case in meditation). The power of the unconscious is such that if a hypnotized man is given the suggestion that he has just eaten meat, he will begin to secrete digestive enzymes. If a forty-year-old woman is told to regress to an infant of three months, she will exhibit Babinsky reaction (fan the toes) when tickled on the sole of the foot. When subjects in deep hypnosis are made to hold a pencil and given the suggestion that it is a pair of red-hot tongs, they will actually exhibit symptoms of a burn. Subjects may also have very powerful audial and visual hallucinations. All these phenomena are simply parts of a dramatic monologue enacted by the powerful unconscious.

The power of the individual unconscious can also be seen in posthypnotic suggestion. While the individual is in a deep state of hypnosis, a hypnotist may implant a suggestion such as, "When I clap three times, open your eyes. When I then clap once, open the window. You will not be able to remember what I have said, but will do what I have told you." Later, when the hypnotist claps once, subjects will behave exactly as directed. When asked why they opened the window, they will reply that, for example, "It's so hot in here, I thought I should open it," even if it isn't really hot at all. This implies that individual unconsciousness, when it gains or is given control of thought and action, has the power to disregard reality totally, and that an explanation of the action afforded by consciousness is basically groundless.

Hypnotic-type states are often produced in the initial stages of meditation practice. The observer (witness consciousness), which is supposed to watch detachedly as contents pour forth from the unconscious, is not strong enough to maintain itself among the panoply of subjective experience pouring out of the unconscious and disappears; the individual again becomes attached to the contents of the individual unconscious and identifies them as being objective experiences of reality. We often encounter this phenomena in people who are strongly attached to a certain religious belief. Out of the individual unconscious comes a striking image of God, Buddha, or Jesus, which is heralded as a revelation. Sometimes, what has actually happened is that a memory of a statue of Buddha or a picture of

Jesus, stored away in the unconscious, has been suddenly released and projected onto the screen of quieted consciousness. In extreme cases, this type of projection produces an unrelenting obsession with the unconscious element. Rather than disposing of the formerly repressed element, the individual becomes completely dominated by it and loses touch with reality. People in the midst of this type of psychological event are apt to misinterpret the intrusion of their own unconscious as being true communication with a higher power, and may declare that they have been receiving messages directly from God, et cetera. They tend to become fanatically narrow-minded. It is taught by many sources that actual unification with higher states of consciousness, in contrast, brings with it an expansive attitude of compassion and tolerance.

As the method to avoid getting trapped by the power of the unconscious, meditation techniques universally demand the cultivation of detachment. Instructions stress that meditators must not identify with the contents of the unconscious as they are being released, but rather watch them impassively. Many different spiritual teachings assert that it does not work to try to repress these contents, because once they have been released, the unconscious does not want them back; it resists, rejecting the phantoms. The situation can be likened to trying to push a ball underwater; trying to suppress the subject matter of the unconscious only makes it rise up all the more forcefully. Therefore, mystical traditions teach that one must let the contents escape without becoming involved, without becoming possessed by the energy of the unconscious, no matter how strong or persuasive it may be.

Once the contents have been released and the individual unconscious is quieted, unification with higher realms of being may begin to take place. The personal, sense-dependent construct of reality is temporarily suspended. Thoughts and impressions not born of one's own personal experience may enter the field of awareness. A very real connection with the outside world is established, a connection not dependent on sensory apparatus. It is in this state that phenomena such as ESP and PK begin to occur.

At first, however, objectively valid experiences of nonsensory dimensions may be confused with residual projections from the individual unconscious. The mystical traditions all counsel that this is the time when a competent teacher or guide—someone with the vision to realize the difference between illusion and reality—is needed most, to help the seeker learn to distinguish between subjective creation and objective fact. This process is gradually being facilitated as increasingly scientific criteria are being created.

A major characteristic of this stage of spiritual evolution is that it is often a period of general disturbance, because the seeker is coming into direct contact with manifestations of being that are quite alien to anything previously known. People identify sense-dependent consciousness with their "selves," and when it is suspended, the annihilation of individuality, fear of death, and the heavy stress caused by clinging to the dissolving personality may produce psychotic-type symptoms. The mind and body, which have been functioning within the limits of the physical plane, become unstable when the familiar framework is broken down. However, in most cases, the instability will soon disappear as a new balance is established.

True psychosis also presents a situation in which the normal balance between the conscious and unconscious sides of personality has been altered. However, in psychosis subjects are unable to exercise any voluntary control: the framework of the personality has broken down to the point where the unconscious is able to make continual and spontaneous appearances in the field of awareness. Deluded subjects passively receive whatever comes into their heads; they have completely lost the ability to distinguish between figments of the imagination and reality.

The situation is further complicated by the fact that when the balance of mind is severely disturbed, as in the case of schizophrenia, perceptions may intrude from realms beyond the individual shell; thus, some of the information received by the schizophrenic may indeed be a true statement of higher objective reality transmitted from nonsensory realms of being. Unfortunately, however, the wheat can not be separated from the chaff by present methods of psychological testing, and in such cases the suitable therapy is often difficult to determine. Moreover, the nonvolitional manner in which a breakdown occurs in the case of schizophrenia lacks the implicit safeguards of the ordered progression leading to a breakthrough which is advocated by the mystical traditions. When the principles underlying the evolution of consciousness are recognized, the process, though extremely difficult, may be accelerated intentionally. Anyone, no matter how healthy, who is successful in the practice of a spiritual discipline will most likely undergo a period of psychological and/or physical instability as a matter of course, since the body and mind must change and adjust to their new relationship to the nonsensory dimensions of existence. This is a natural function, and no cause for alarm.

This idea fits in well with chakra theory. Each chakra is alleged to control a specific area of ordinary unconscious content, just as they are seen to oversee a corresponding system of physiological function. As the ordinary arena of each chakra's unconscious contents is made conscious through the process of spiritual evolution, energy is released which will affect the functioning of the systems related to that chakra. Therefore, different types of imbalance noted in a given subject can help us to determine which of the chakras is in the process of being awakened.

If the instability in an individual is caused by the initial breakthrough into nonsensory forms of consciousness rather than by simple pathological factors, there should be corresponding evidence of the ability to use nonsensory modes of perception and activity. The type of evidence found can be used as a guideline to help an individual determine where he/she stands vis-a-vis the meditative discipline itself.

To help clarify the differences between individuals in the process of breakthrough and those suffering from breakdown, I decided to compare bioelectric examinations of patients clinically diagnosed as psychotic with similar examinations of group A and B subjects. In general, I found that group A subjects can easily be distinguished from individuals diagnosed as being mentally unbalanced, whereas many group B subjects initially show distinct similarities to the deranged. The polygraph research, for example, showed a wider span of autonomic control in group A subjects, but a span which is balanced and consistent. In an ordinary, healthy subject, the baseline of the plethysmograph is rather flat, indicating that the workings of the autonomic nerves are constant, not unstable or hypersensitive. The plethysmographs of many group A subjects evidence great fluctuation in the baseline, indicating hypersensitive activity. But a distinct rhythm to the fluctuation is also evident, indicating a new type of stability. Whereas these patterns were quite consistent for hundreds of normal and group A subjects tested, I was unable to find any pattern in the case of mental patients. There appears to be no fixed tendency in the autonomic nervous functions of the mentally ill.

A series of VCR studies produced findings which corroborated the above results. After stimulating the VCR points electrically and measuring the resultant skin resistance, I began to notice some general tendencies: Control subjects showed little difference in skin current values before and after stimulation, which suggests that in their case the autonomic nervous function that controls the internal organs is stable. In the case of many subjects from groups A and B, on the other hand, I found a significant difference between pre- and post-stimulation data. Often this difference would show up more markedly in one organ, and the same three organs (heart, stomach, kidney) were found to be most unstable. The mentally unbalanced subjects also showed a remarkable difference in their pre- and post-stimulation data, but this difference did not appear to be connected to any specific organs. Rather, the differences seemed to indicate an imbalance in the entire autonomic nervous system, similar to the type of imbalance seen when the hormone secretion system is not working properly.

Physiological testing equipment reveals many similarities in the organic function of both group B and mentally unstable subjects, but is not designed to delineate the differences. The AMI and chakra machines, however, are potentially more effective tools with which to investigate a subject's condition. Initial findings indicate, for example, that an individual with an active chakra tends to produce AMI readings which reflect unusual excess of energy in the meridians connected to that chakra. People in whom empirical evidence suggests an awakened manipura chakra almost always show an excess of energy in the stomach and spleen meridians, whereas anahata activity is similarly reflected in the heart and heart constrictor meridians. Emotionally or mentally disturbed individuals do not manifest this pattern. Another finding is that the average of the BP values of each meridian (an indication of the overall metabolic state of the organism) is very high for those with apparent chakra activity whereas it is very low in psychologically unbalanced subjects.

A certain degree of instability is a normal function of spiritual evolution. Occasionally, however, individuals will encounter real difficulties when they attempt to expand their consciousness without proper instruction. The mechanism of kundalini awakening, for example, sometimes presents problems if the principles of evolution are misunderstood or acceleration of evolution is triggered prematurely. One of the most complete descriptions of kundalini is found in Gopi Krishna's work *Kundalini—The Evolutionary Energy in Man.*[3] Gopi Krishna began meditating at age seventeen, practicing daily while leading the life of a civil servant and family man. One morning seventeen years later, as he was practicing concentration at the crown of his head, he felt a strange sensation at the base of his spine. He describes the sensation: "Suddenly, with a roar like that of a waterfall, I felt a stream of liquid light entering my brain through the spinal cord." Thus began a tortuous ordeal which lasted for many years, during which time Gopi Krishna was driven nearly crazy by the strength of the fiery energy unleashed within him and came close to death. The energy brought with it moments of bliss and transcendence, but also a very painful neurophysiological transformation, a complete change in the way his body, and later his mind, functioned.

Not everyone suffers so terribly with the awakening of kundalini. This pain can be circumvented, particularly if meditators are prepared for the phenomenon. Gopi Krishna later realized that his problem was caused by the fact that the kundalini traveled up the wrong vital pathway: traditional empirical reports state that kundalini is supposed to travel up the central spinal channel, the sushumna, but in his case it was unleashed into the pingala. He writes:

> The hot blast coursing through my nerve and brain cells would have undoubtedly led to death but for the miraculous intervention at the last minute [the cooling power of ida was activated]. Later on, my suffering was probably due, firstly, to the damage already sustained by my nervous system; secondly, to the fact that I was entirely uninitiated into the mystery, and thirdly and mainly, to the circumstance that my body, although developed above the average in muscular strength, was not sufficiently developed internally to withstand with impunity the sudden onrush of a much more dynamic and potent life energy than that to which the average human body is normally accustomed.[4]

Gopi Krishna is convinced that some people locked away in mental institutions as clinically insane actually suffer from precipitously awakened kundalini. He sees a great need for thorough scientific investigation into the specific mechanisms of this experience, and I share his sense of urgency. I feel that the AMI may be very helpful in determining the safest types of spiritual practice for a given individual. We can partially determine, for example, which chakra, if any, is naturally more active than the others and ascribe practice accordingly. The AMI is also able to monitor the balance of ki in the body, and any imbalance can be repeatedly checked and corrected by actively employing specific types of pranayama and concentration techniques. If someone is found to have an excess of ki in the upper part of the body, concentration on the swadhistana chakra and prolonged abdominal breathing will remedy the situation. In a similar fashion, numerous techniques may be employed in conjunction with AMI readings to insure the maintenance of balance within the system throughout a period of ascetic discipline. Continued research in this direction should aid in the development of guidelines which will greatly reduce the risks sometimes encountered on the spiritual path.

Two American researchers, Dr. Lee Sanella and Itzhak Bentov, have tried to relate kundalini symptoms to western physiology. In his book *Kundalini—Psychosis or Transcendence?*,[5] Dr. Sanella clearly states that the physiological syndromes he reports are merely reflections of a process of consciousness. Investigating many reported cases of kundalini awakening, both Sanella and Bentov trace common patterns of physiological symptoms, attempting to develop criteria by which nonpsychic medical practitioners could diagnose the condition correctly.

More research is needed into the symptoms that attend changes in consciousness as it undergoes the natural process of evolution. It is fairly certain now that these changes occur—that nonsensory states of consciousness do exist, but we need to know more about the specific mechanisms involved, not only to add to the general store of knowledge, but to help individuals who desire to explore the potential forms of consciousness which lie within themselves. I hope that my own work will contribute to a greater understanding of these mechanisms and thereby make the path of breakthrough a little easier to follow.

AFTERWORD

MY FIRST meeting and association with Dr. Hiroshi Motoyama was in the Philippines in the year 1966, where we went to study the highly controversial Psychic Surgery taking place there, and which resulted in my book *Wonder Healers of the Philippines*. A close friendship developed which in 1972 brought Dr. Motoyama to lecture to our Body, Mind and Spirit Healing Workshop at the Arlington Hotel, in Hot Springs, Arkansas, and took me in return to Tokyo, Japan, to speak for him in 1977, at his International Metaphysical Conference.

I was impressed, from the first, with Dr. Motoyama's advanced scientific background, the thoroughness of his research methods, and his announced objective to find a way to unite Science and Religion in their conceptual acceptance of physical and nonphysical dimensions of existence. On reading this, his new book, *Science and the Evolution of Consciousness*, I would say he has come close to hitting the target. While Dr. Motoyama modestly states that his findings will require further corroboration, and merit further investigation, his achievement is no less monumental.

As an aid to his research on and treatment of subjects, he has invented and employed what he calls an "AMI Machine" which correctly diagnoses the conditions of different organs, all in a few minutes time.

In addition, he has developed a "chakra machine," a sensitized device capable of measuring the potential of the electrical field surrounding a subject's body without the need to attach any electrodes. It functions inside a lead and copper shield. An electrode placed in front of different areas in the body registers changes occurring as feeling and emotional attitudes change. The AMI and Chakra Machine can give an accurate read-out on what is happening to the body chemistry.

Dr. Motoyama has much to say about the chakras and their functions. He emphasizes the danger of awakening the kundalini too quickly, thus causing pain and wrong vital patterns to develop, but he also points out the spiritual rewards which can ensue if mental and emotional control is exercised and maintained.

The contents of this book require profound study. Dr. Motoyama has added the ancient Chinese practice of acupuncture to his own instrumentation, stressing the value of knowing how to contact the various meridians having to do with the autonomic nervous system, which act like telegraph wires in stimulating different organs and glands, thus restoring their normal functioning.

The full name of Dr. Motoyama's diagnosing machine, is the "Apparatus for Measuring the Functioning of the Meridians and Corresponding Internal Organs." This computerized apparatus can diagnose imbalance anywhere in the physiological organism in a matter of minutes. It has 28 electrodes, to be attached to 28 meridians and points on fingers and toes, with 3 volts, D.C. applied sequentially to measure skin currents. In his experimentations, Dr. Motoyama has measured over 5,000 subjects, establishing normality and abnormality values. These tests are programmed into computer read-outs. The Motoyama AMI Machine is being adopted by an increasing number of hospitals and clinics in different parts of the world.

Says Dr. Motoyama, "We have proved conclusively that the human is more than a body and a limited intellect. He possesses nonphysical properties and energies that are still largely unknown and uncharted by Science." Man has three bodies—the physical, astral, and causal—which overlie and interpenetrate each other. Each level of Being is sustained by the energy in that dimension. More refined matter is to be found in the higher dimensions. Physical tools can only measure physical objects. Modern science is therefore unable to detect or evaluate nonphysical energies. The forces we are dealing with are so subtle and so far removed from what we regard as the material world that they will completely revolutionize physics, mediums, and philosophy and bring about a profound change in humanity's perception of reality. Each higher dimension appears to be a world in itself, still largely beyond our comprehension and ability to adequately evaluate.

Dr. Motoyama raises basic questions as to the nature of the connection between mind and matter—the non-physical and the physical. It is his observation that we are ordinarily unaware of the higher dimensions of our being; that all Nature maintains a balance between the positive and negative forces; that there are what the

Chinese call yin and yang aspects of energy which can be influenced by the manipulation of the meridians or nerve centers which influence different organs of the body; that Consciousness can directly exert a psychokinetic effect upon the external environment; that the astral body contains all dreams and emotions and attracts what it desires; that feeling provides the power behind thought.

This discovery and confirmation of the power of feeling, as demonstrated by Dr. Motoyama's experiments, dates back to the now classically regarded long distance telepathic tests conducted by Sir Hubert Wilkins and myself in the fall of 1937 and spring of 1938, when Wilkins was in the Far North searching for lost Russian fliers, and I, acting as receiver, was situated in my study in New York City, separated by two to three thousand miles. We proved conclusively, as reported in our book, *Thoughts Through Space*, that the experiences Wilkins had which affected him the most emotionally I was able to perceive the easiest. It was as though feeling generated the power behind thought: the more mental images were associated with feeling, the more readily they were impressed upon my mind as I fixed its attention on Wilkins at our appointed times.

In that period, as it happened, no other means of communication was possible. The sunspot conditions were so bad that Wilkins, near the North Pole, could only maintain spasmodic radio contact with the New York Times short wave radio station. As it turned out, checked against Wilkins' diary and log, I had more accurate day-to-day knowledge of Wilkins' activities than was possible through mechanical facilities.

Perhaps, with all of Dr. Motoyama's exhaustive research, his re-confirmation of the power of feeling with respect to the transmission and reception of thought, may prove among the most significant. Feeling, I have long contended, will be found to be the universal language—the key to communication between all forms of life.

Dr. Motoyama has noted the effect on the solar plexus of an emotional disturbance. It was my observation, during the Wilkins experiments, and this has been true of all my telepathic research since that time, that mental imagery impressions seem to strike the brain region first and then become "grounded" in the solar plexus, or what some scientists call "the second brain." When I would get a "gut" feeling in my solar plexus almost simultaneous with a thought impulse entering my mind, I usually accepted this as the signal that I had received, or was receiving, a genuine impression.

The question: Is Dr. Motoyama's AMI Machine sensitively picking up and recording, through electrical impulses, the feelings generated by the different organs of the body and expressed magnetically? We know the effects of negative and positive charges in matters of mental and physical health or ill health. Are we getting closer to a translation of human feelings in the form of interpretive instrumentation?

With respect to the problem of healing, Dr. Motoyama has observed that when the Ego of the individual is permitted to enter in, it seems to create a blockage or an obstructive influence so that the healing energy flow from the mind of the healer, which might otherwise be transmitted to a person in need of a healing treatment, cannot get through.

"You must empty the mind of a sense of the personality as much as is possible," says Dr. Motoyama. I have found this to be true in my own personal experimentation. The more the awareness of Self can be removed from an attempt to reach a target, the more the awareness of the target itself can be achieved.

It has been found that those who practice different forms of meditation faithfully, seem to develop, in time, greater intuitive and ESP ability, just as a muscle of the body, regularly exercised, grows in strength and usefulness. People with close emotional ties can sense each other's feelings and thoughts more readily. There is a change in breathing and heart action often noticed during meditation or efforts at concentration. But it is evident that telepathy does not depend upon physical energy. Sensitives placed in a Faraday cage, shielded from all external electromagnetic influences, are not inhibited from receiving thought impressions. Consciousness is apparently an independent entity which functions through the brain as an instrument. Dr. Motoyama's experiments have shown that mind can influence the working of the body as well as another person's body. The functioning of mind can be inter-related, with emotion always playing an important role.

Dr. Motoyama's "Science and the Evolution of Consciousness" has opened wide the door to new scientific and spiritual speculation. He has given us, in a sense, a textbook for the future. I salute him for his courage, his integrity, and the further light he has shed, through his advanced research, on the Unknown areas of body and mind.

Harold Sherman
Founder, ESP Research
Associates Foundation

APPENDIX A

Chakras and the Autonomic Nervous System

Another method was employed to explore further the relationship between the anahata (heart) chakra and the corresponding autonomic nerves controlling the heart. Variations of blood pressure for the three classes were determined by measurements taken before and after a stimulus was applied to the left head zone for the heart. Groups A and B displayed decreases in blood pressure much greater than those of group C. A decrease or lack of variation in blood pressure would signify predominance or strain of the parasympathetic nerves, whereas an increase would mean sympathetic nervous system predominance. Therefore, the results of this investigation agree with those obtained through measurement of skin current variation and the questionnaires on disease. A predominance of the vagus (parasympathetic) nerve would be expected to weaken those cardiac functions normally accelerated by sympathetic nerves, because the parasympathetic nerves serve an inhibitory function, and such abnormalities of the heart are in fact frequently found in subjects from groups A and B who report an awakened anahata chakra.

Another way of correlating chakra awakening with autonomic nervous activity is to examine vertebral dislocations and distortions. Both chakra awakening and vertebral dislocations have been observed to coincide with excitement or tension in the autonomic system and some abnormal functioning in the organs or areas controlled by the autonomic nerves originating from the related spinal segments. In addition, points close to the dislocated vertebra often become sensitive and sore. It seems that the dislocated vertebrae put a strain on spinal nerves which eventually brings about abnormality in the functioning of corresponding organs, while points of sensitivity arise from friction or pressure against the ligaments, muscles, and other tissues attached to the concerned vertebrae.

When the first through fourth thoracic vertebrae were examined for the three classes of experimental subjects, it was found that group A showed the highest frequency of vertebral dislocation. In the case of the fourth vertebra, where the plexus controlling heart and bronchi issues is situated, the rate of dislocation was highest among group A subjects. This correlation seems to agree with the unusual incidence of reported anahata (heart) chakra awakening among those belonging to group A. Vertebral dislocation was again found to be predominant in group A when the fifth through the twelfth thoracic vertebrae were examined. Nerves emanate from this vertebral segment to the solar plexus and inervate the abdominal organs, so we suspect some functional correspondence to the manipura chakra.

Finally, I observed the frequency of dislocations from the first lumbar to the fourth sacral vertebrae, whose nerves spread out to form autonomic ganglia governing the function of the large intestines and genitourinary organs. These ganglia appear to correspond on the whole to the swadhistana chakra. Again, group A exhibited the highest frequency of dislocation, and particularly high figures for A and B were shown for the fourth and fifth lumbar vertebrae.

Assuming that vertebral dislocation would result in some excitement of the related autonomic nerves, these results would indicate that (1) the autonomic nervous system of group A subjects is abnormally excited; (2) the particularly notable abnormality recognized for these subjects in the region from the fifth to the twelfth thoracic vertebrae supports their reports of activity in the area of the manipura chakra and also a correspondence between the solar plexus and this chakra, since autonomic nerves connect these vertebrae with the solar plexus. Similar inferences can be made regarding the anahata–thoracic vertebrae and the swadhistana–lumbar vertebrae relationships.

APPENDIX B

Meridian Test 2

As in the experiment described in the text, the triple heater meridian was chosen as the object to be tested. The experiment was first performed on three subjects, by monitoring the galvanic skin potential (GSP) at the following points on the triple heater meridian: *seiketsu* point (*kansho*), *tensei*, *tenryo*, *ketsubon*, *danchu*, *sekimon* (triple heater *bo* point), and *yu* point. A point on the right palm was also monitored, to observe the possible reaction of the sympathetic nervous system. In the first series of the experiment (series A), the electric stimulus was applied to the seiketsu point in the first test while all other points were being monitored; similarly, in the second test the yu-point was stimulated while the other points were being monitored. Later, in series B, seven more subjects were tested in exactly the same way, except that the large intestine yu point was also monitored. This control was to allow comparison between the triple heater yu point, which is assumed to have a close functional connection to the triple heater meridian, and the large intestine yu point, which is supposedly not closely connected to the triple heater meridian.

Perception of the given stimulus was primarily a subjective process: for example, one subject did not begin to feel pain until the stimulus was increased up to 75Vdc, 90msec duration, whereas another subject felt intense pain with a stimulus as low as 20Vdc of the same duration. Therefore, it is impossible to derive an objective criterion to define the effective intensity of a given stimulus. The use of words such as "strong stimulus," "moderate stimulus," "weak stimulus," will necessarily remain somewhat vague. However, this subjective ambiguity does not effect the results of the experiment.

Results of stimulation of the seiketsu point

Series A

When a sufficiently strong stimulus was given, GSP was observed at all the points monitored. One of the subjects showed recurrent GSP reaction at intervals of 5 to 18 seconds, first at all the points and then later at the triple heater yu point, sekimon, and danchu only. With a stimulus of moderate intensity, all the subjects showed GSP at the triple heater yu point. Two subjects showed reaction at the sekimon as well.

Series B

A strong stimulus was given to four subjects and a moderate stimulus to the remaining three subjects. Of the four subjects given strong stimuli, three showed GSP at all the points monitored, and one showed GSP at all the points except the large intestine yu point. Of the three subjects given moderate stimuli, all showed GSP at the sekimon, the right palm, and the large intestine yu point. Two showed the reaction at the triple heater yu point as well.

These results may be summarized as follows:

If the given stimulus is sufficiently strong, GSP is generated at all the points monitored. This GSP reaction implies that for a strong stimulus the sympathetic nerves of the whole body are simultaneously excited. However, when a moderate stimulus is applied (that is, when a given stimulus is neither so weak as to show no reaction anywhere nor so strong as to show reaction everywhere), five subjects out of the total ten subjects showed reaction at the sekimon and triple heater yu point, and three showed reaction at the right palm.

In Series B, three subjects out of seven showed reactions at the large intestine yu point as well as at the sekimon and the triple heater yu point, when moderate stimulus was given.

From the neurophysiological standpoint, the seiketsu point of the triple heater meridian—located near the ulnar base of the fourth fingernail—belongs to the segment C_8 of the spinal nervous dermatome and T_1-T_4 of the sympathetic nervous dermatome. On the other hand, the triple heater yu point and the sekimon belong to L_1-L_2 and T_{11}-T_{12} respectively. Thus, there is no close connection via the nervous system between the seiketsu point and the triple heater yu point or the sekimon. Furthermore, considering the fact that no GSP was generated at the other points monitored, which are distance-wise closer to the seiketsu point (the point of stimulation), than the triple heater yu point or the sekimon, one must conclude that the change in the electric field incurred by the electric stimulation was, in fact, negligible. Thus, how can the fact that five of the ten subjects showed reaction at the sekimon and the triple heater yu point be explained? Acupuncture theory states that there is a close connection between the seiketsu point, the yu point, and the bo point (sekimon, in the case of the triple heater meridian) of a given meridian. The results of this experiment become plausible if we assume the existence of the meridian system.

Stimulation of the triple heater yu point

Series A

When a strong stimulus was given, all the subjects showed GSP at all the points monitored, and one of them showed a large recurrent GSP at the seiketsu point about 4 seconds after the stimulation. When a moderate stimulus was given, one of the subjects showed a reaction at the sekimon and another at the seiketsu point.

Series B

Of the seven subjects, four were given strong stimuli and all showed a reaction at all the points monitored. Weak stimulus was given to the remaining three. No reaction occurred at any points in any of the subjects. When a moderate stimulus was given to the chosen four subjects, all showed GSP at the seiketsu point as well as at the palm. Two out of the four showed additional reaction at the sekimon, and one also at the chukan.

The above results may be summarized as follows:

When a sufficiently strong stimulus was given, a GSP reaction was observed at all the points monitored. When a moderate stimulus was applied, however, five out of the six subjects showed a reaction at the seiketsu point; four showed a reaction at the right palm, and three at the sekimon.

These results suggest the presence of a strong correlation between the triple heater yu point and the seiketsu point. As I have already pointed out, there is no close neurophysiological connection between these two points. Therefore, it again seems plausible to posit the existence of the meridian system as an explanation of the noted phenomena.

The GSP that was observed at the right palm in four out of six subjects may have been induced in the following manner: The sympathetic nerves controlling the sweat glands in the palm were excited by the stimulation of the triple heater yu point, triggering a depolarization in the plasma membrane which resulted in the generation of GSP.

As for the triple heater yu point and the sekimon, they belong to the segments L_1-L_2 and T_{11}-T_{12} respectively. Therefore, these two points are not closely connected, neurologically. However, according to acupuncture theory, yin-yang relationship exists between the yu point (located on the back) and the bo point (located in the front); therefore, it is not surprising to find a relatively strong correlation between these points of the triple heater meridian. The fact that three out of the six subjects showed a reaction at the sekimon provides further evidence suggesting that the meridian system does indeed exist.

FOOTNOTES

INTRODUCTION

[1] D. E. Woolridge, *The Machinery of the Brain* (McGraw-Hill Book Co., Inc., New York, 1963).
[2] R. D. Becker and D. G. Murray, "The Electrical Control System Regulating Fracture Healing in Amphibians," *Clin. Orthop. and Rel. Res. 75*, 169 (1970).
[3] D. Hawkins and L. Pauling, *Orthomolecular Psychiatry* (W. H. Freeman and Co., San Francisco, 1973).
[4] M. Reichmanis, A. A. Marino and R. O. Becker, "Electrical Correlates of Acupuncture Points," To be published in *IEEE, Trans. on Biomed. Eng.*, 1975.
[5] W. A. Tiller, "Some Physical Network Characteristics of Acupuncture Points and Meridians," Academy of Parapsychology and Medicine, Acupuncture Symposium, Stanford University, California, June 1972.
[6] V. Adamenko, private communication, Soviet Union, September 1971.
[7] E. Green, "Biofeedback for Mind-Body Self Regulation: Healing and Creativity," Academy of Parapsychology and Medicine, Varieties of Healing Experience Symposium, De Anza College, California, Oct. 1971.
[8] B. B. Brown, *New Mind, New Body—Biofeedback: New Directions for the Mind* (Harper and Row, New York, 1975).
[9] Sister Justa-Smith, "The Influence of Enzyme Growth by the 'Laying-On-of-Hands,'" Academy of Parapsychology and Medicine, Dimensions of Healing Symposium Proceedings, Los Altos, California, 1972.
[10] R. N. Miller, P. B. Reinhart and A. Kern, *Thought as Energy*, ed. W. Kinnear (Science of Mind Publications, Los Angeles, California, 1975).
[11] A. Puharich, "The Search For a Common Denominator in Medicine and Healing," Academy of Parapsychology and Medicine, Dimensions of Healing Symposium Proceedings, Los Altos, California, 1972.
[12] W. A. Tiller, "Some Psychoenergetic Devices," *A.R.E. Journal 7*, 81 (1972). (A.R.E. = Association For Research and Enlightenment, Virginia Beach, Va.)
[13] H. Motoyama, "The Mechanism by which PSI-Ability Manifests Itself," *Impact of Science on Society 24*, 321 (1974).
[14] P. Tompkins and C. Bird, *The Secret Life of Plants* (Harper and Row, New York, 1972).
[15] C. Backster, "Evidence of Primary Perception in Plant Life," *International Journal of Parapsychology, 10*:6 (Winter, 1968).
[16] M. Vogel, "Man-Plant Communication," in *Psychic Exploration*, editors E. D. Mitchell and J. White (G. P. Putnam's Sons, New York, 1974).
[17] K. A. Horowitz, D. C. Lewis and E. L. Gasteiger, "Plant 'Primary Perception': Electrophysiological Unresponsiveness to Brine Shrimp Killing," *Science 189*, 478 (1975).
[18] W. O. Schumann, "Uber die Strahlungslosen Eigenschwing ungen einer leitenden Kugel die von einer Luftschicht und einer Ionospharenhulle umgeben 1st," *Z. Naturforsch 7a*, 149 (1952).
[19] H. Konig, "Biological Effects of Extremely Low Frequency (ELF) Electrical Phenomena in the Atmosphere," *J. Interdiscipl. Cycle Res. 2*, no. 3, 317 (1971).
[20] W. A. Tiller, "A.R.E. Fact Finding Trip to the Soviet Union," *A.R.E. Journal 7*, 68 (1972).
[21] I. Swann, *To Kiss Earth Good-bye* (Hawthorne Books, Inc., New York, 1975).
[22] H. Forwald, *Mind, Matter and Gravitation* (Parapsychology Foundation, New York, 1969).
[23] J. Taylor, *Superminds* (Macmillan, London, 1975).
[24] R. Targ and H. Puthoff, "Information Transmission Under Conditions of Sensory Shielding," *Nature 252*, 602 (1974).
[25] L. Day and G. de la Warr, *New Worlds Beyond the Atom* (Vincent Stuart, London, 1958).
[26] L. Day and G. de la Warr, *Matter in the Making* (Vincent Stuart, London, 1966).
[27] W. A. Tiller, "Radionics, Radiesthesia and Physics," Academy of Parapsychology and Medicine, The Varieties of Healing Experience Symposium Proceedings, Los Altos, California, 1972.
[28] Z. V. Harvalik, "Sensitivity Tests on a Dowser Exposed to Artificial D.C. Magnetic Fields," *American Dowser 13*:3, 85 (1973). Ibid 14:1, 4 (1974).
[29] W. A. Tiller and W. Cook, "Psychoenergetic Field Studies Using a Biomechanical Transducer," Proceedings of A.R.E. Medical Symposium on New Horizons in Healing, Phoenix, Arizona, January 1974.
[30] S. Ostrander and L. Schroeder, *Psychic Discoveries Behind the Iron Curtain* (Prentice Hall, New York, 1970).
[31] S. Karagulla, *Breakthrough to Creativity* (De Vorss and Co., Inc., Los Angeles, 1967).
[32] H. Puthoff and R. Targ, "Psychic Research and Modern Physics," in *Psychic Explorations*, editors E. D. Mitchell and J. White (Putnam's Sons, New York, 1974).
[33] G. Milne, *The Pyramid Guide 18*, 5 (1975).

CHAPTER 1

[1] Fritjof Capra, *The Tao of Physics* (Shambhala: Colorado, 1975), p. 150.
[2] Gopi Krishna, *The Biological Basis of Religion and Genius* (Harper and Row: New York, 1971), p. 77.

CHAPTER 2

[1] Evelyn Underhill, *Mysticism* (Dutton: New York, 1961).
[2] William James, *The Varieties of Religious Experience* (Longmans, Green & Co.: New York, 1928), p. 388.
[3] Patanjali, *Yoga Sutras* (Poona, 1904; 2nd edition 1919) III, 2.
[4] Jean Gouillard, *Petite Philocalie de la priere du coeur* (Paris, 1953), p. 216.
[5] As quoted in Underhill, op. cit., p. 319.
[6] Mircea Eliade, *Yoga: Immortality and Freedom* (Bollingen Series: Princeton, 1969), p. 99.

CHAPTER 3

[1] The scientist, W. E. Cox, conducted a study which suggests such premonitions of accident do occur significantly. He did a statistical analysis on the number of passengers aboard twenty-eight railroad trains which met with accidents, finding that the number was significantly less than could be expected according to the number of passengers riding the same trains in the same time period on other days.
[2] Yogi Satyanarayana, "Kundalini Yoga," *Religion and Parapsychology*, No. 14, Tokyo, 1976.
[3] Louisa Rhine, *ESP in Life and Lab: Tracing Hidden Channels* (Macmillan: New York, 1967), p. 168.
[4] *Proceedings of the Parapsychological Institute of the State University of Utrecht*, Vol. 3, 1965.

CHAPTER 4

[1] Douglas E. Dean, "The Plethysmograph as an Indicator of ESP," *Journal SPR*, 41, 1962; pp. 351–353.
[2] Charles T. Tart, "Possible Physiological Correlates of Psi Cognition," *International Journal of Parapsychology*, 5, 1963; pp. 375–386.
[3] Delcorte Press/Seymour Lawrence, 1977; p. 58.
[4] DeVorss & Co., Los Angeles, 1967.

CHAPTER 5

[1] The Theosophical Publishing House, Illinois, 1971; pp. 7–8.
[2] Ibid., p. 3.

CHAPTER 6

[1] Personal communication.
[2] Y. Manaka and I. Urquhart, *The Layman's Guide to Acupuncture* (Weatherhill, New York, 1972), p. 21.
[3] Yoshio Nagahama and Masaaki Maruyama, *Studies on Keiraku* (Kyorinshoin Co., Ltd.).
[4] It seems probable that there is a direct relationship between these finger and toe points and the points detected by Kirlian photography, a process in which a high-frequency voltage applied to the body causes an illumination which can be photographed. The fingertip points which shine so brilliantly in Kirlian photographs particularly in the case of psychic healers, may be physical representations of the ki energy emanating through the fingertips.
[5] For a more precise discussion of the development and mechanics of this apparatus, please refer to my book, *How to Measure and Diagnose the Function of Meridians and the Corresponding Internal Organs* (Tokyo, Japan: The Institute of Religion and Psychology, 1974).

CHAPTER 7

[1] R. D. Laing, *The Politics of Experience*, (Pantheon: New York, 1967).
[2] C. Naranjo and R. Ornsteir, *On the Psychology of Meditation* (Viking: New York, 1971), p. 111.
[3] Gopi Krishna, *Kundalini: The Evolutionary Energy in Man* (Shambhala: Berkeley, 1970), p. 12.
[4] Ibid., p. 137.
[5] Lee Sanella, *Kundalini: Psychosis or Transcendence* (Henry S. Dakin: San Francisco, 1967).